Marketing of Engineering Consultancy Services:
A Global Perspective

Marketing of Engineering Consultancy Services:
A Global Perspective

Dr. Pradeep Kumar Chatterjee

B.E. (Mech. Engg.), M.B.A. (Mktg), Ph. D (Mktg. Mgt), A.M.T, F.I.E, M.A.I.M.A, M.I.S.C.A, M.N.I.P.M, S.M.C.S.I, M.L.M.I.I, Chartered Engineer (India), M.A.S.M.E. (USA), M.A.M.A (USA)

Jt. General Manager (Procurement Services) MECON Limited, Ranchi (India)

And

Honorary Adjunct Faculty and Member Academic Committee, Department of Management, Birla Institute of Technology, Mesra, Ranchi (India)

Partho Pratim Chatterjee

B.E. (Metallurgy), A.M.I.E, M.I.S.C.A, M.A.S.M.E (USA), S.M. A. I. S.T (USA), S.M.T.M.S (USA)

The Technical Manager's Survival Guides

ASME PRESS

Library of Congress Cataloging-in-Publication Data

Names: Chatterjee, Pradeep Kumar, author. | Chatterjee, Partho Pratim, author.
Title: Marketing of engineering consultancy services : a global perspective / Pradeep Kumar Chatterjee and Partho Pratim Chatterjee.
Description: New York : ASME Press, [2019] | Includes bibliographical references and index.
Identifiers: LCCN 2018045162 | ISBN 9780791861837
Subjects: LCSH: Consulting engineers. | Consulting firms--Marketing. | Consulting firms--Management.
Classification: LCC TA216 .C495 2019 | DDC 620.0068/8--dc23 LC record available at https://lccn.loc.gov/2018045162

Dedication

Dedicated to my beloved mother Late Shibani Chatterjee, (a retired school teacher) for being the main spring of inspiration, for being there at every stage of arduous journey and who emotionally moulded me and taught me anything was possible and whose influence on my academic, professional and personal life has been profound, prolific and indelible.

"If my mind can conceive it, and my heart can believe it—then I can achieve it."

Muhammad Ali

Table of Contents

List of Tables

List of Figures

Foreword

Marketing of Engineering Consultancy Services has always been a complex and multi-disciplinary subject, more so in a world of interconnected economies. The Service Dominant (S-D) logic argues that all exchanges can be viewed in terms of service-for-service exchange. This in itself embodies the ubiquitousness and importance of service marketing.

The cross-border sourcing of services as an integral part of the global value chain gained momentum with enhanced economic globalization. This in turn gained strength from increasingly diminishing relevance of geographical location and distance brought about by technology revolution. We are in the cusp of the creative disruption of Industrial Revolution 4.0 (IR 4.0) where technology is set to change the way we live and work. This technology disruption is spearheaded by Artificial Intelligence (AI), Internet of Things (IoT) and Machine Learning. As the global economy and society are witnessing a tectonic shift to post- industrial information society with knowledge capital as its new resource, marketing of services is concurrently witnessing a paradigm shift. To reiterate in Philip Kotler's own words, "Markets are changing faster than Marketing."

The pace and scale of engineering innovation that mankind has witnessed in last two decades is unprecedented in history. Marketing strategies of engineering consultancy services in this backdrop has seen many new normals, but none being enduring. Mostly so because marketing itself is embedded in a particular time, space and culture and evolving a Universalist strategy has remained elusive.

To keep pace with the global giants in a hypercompetitive engineering consultancy service domain is both challenging and rewarding. We at MECON have developed a unique way of working by harnessing our cultural capital, legacy, tradition and an underlying *gemeinschaft*. This personalized and informal way of working, which permeates the ostensible bureaucratic structure, has been our biggest asset which can be aptly epitomized as the "MECON way" which forms the bedrock of our soft branding.

This book is a unique blend of Dr. P.K. Chatterjee's three decades of illustrious career in engineering service industry, dovetailed with his scholastic academic excellence. It is my firm conviction that this book will surely outlive the present challenges and address the future complexities associated with marketing of engineering consultancy services and cater to the much needed industry-academia linkage.

Goutam Chatterjee
Director (Commercial)
Dated : 06th July, 2018   MECON Limited, Ranchi, India

Preface

My exposure to the engineering consultancy services sector began when I joined MECON Limited as Management Trainee Technical (MTT) in the beginning of the year 1984, an amateur fresh from engineering college posted at Steel Authority of India Limited/ Rourkela Steel Plant, Rourkela, India. MECON Limited [erstwhile Metallurgical and Engineering Consultant (India) Limited] is a public sector engineering and consultancy organization under administrative control of Ministry of Steel, Government of India. During the training period, in a class room session, we were made cognizant of the more than a century old valuable saying of our great leader and Father of the Indian nation Mahatama Gandhi about the customer.

Quote

A customer is the most important visitor in our premises. He is not dependent on us, we are dependent on him. He is not an interruption in our work, he is the purpose of it. He is not an outsider in our business, he is an integral part of it. We are not doing him a favor by serving him, he is doing us a favor by giving us an opportunity to do so

Unquote

Today, when I recollect, I realize that even at that time in an engineering consultancy service organization there was great concern for customers and the critical role the front end employees (engineers), who are the real service providers, play in delighting them. It does not matter that the practice was not documented in Apex Quality Manual (AQM). About the same time, I came across an article titled "Breaking Free from Product Marketing" published in the *Journal of Marketing* (Vol. 41, No. 2, 1977) authored by G. L. Shostack which initiated a discussion on the marketing approach to services. Later Christopher Lovelock, Christian Gronross, A. J. Magrath, A. Parsuraman, Valarie A Zeithaml, Leonard L. Berry, B. H. Booms, M. J. Bitner, Donald Cowell, Adrian Payne, Tatsuya Kimura, Les Carlson, Florian Kraus,

Tillmann Wagner and others carried out a series of research studies in services marketing. These research findings proposed that for marketing services, 4Ps are not enough. Internal marketing, managing quality, client perception, people factor, physical evidence, process of rendering services, pace, customer delight, customer orientation, etc. are very indispensable for service success.

Later, I observed that Management Schools had started responding to the need of industry by offering courses on services marketing not as a choice but as the need of the hour. However, they had no option but to rely on very few books, and that there were practically none in the area of marketing of intellect services like engineering consultancy, management consultancy, etc. This paucity of a proper text book continued to elude the academic/industrial community, was perturbing and that became the driving force to initiate the work on *"Marketing of Engineering Consultancy Services: A Global Perspective."* Though, sufficient time has elapsed and the wait for a proper text book in this area continues to be present and in demand, this proposed text book is a humble and sincere effort from an engineering consultancy practicing professional cum teacher preaching business to business (B2B) marketing like me to bridge this void.

The marketing of engineering consultancy services in the domain of B2B industrial marketing has become more exciting and continues to be challenging for several reasons. The predominant reasons are: intensified competition with wafer thin margin, expanding markets with ongoing globalization, demanding and discerning clients looking for global quality service, fast growing e-commerce applications, rising importance of high-tech businesses, customer-centric and ethical marketing philosophies and increasing emphasis on knowledge and innovation, etc.

There are text books (though few) on engineering consultancy, but this one is different. This book seeks to serve the knowledge needs of both practitioners as well as management students. To be cynosure to both the groups, the book is designed with several relevant contemporary examples. The main strength of this book is its comprehensive coverage at the micro level. Along with the conventional marketing themes like marketing philosophy, marketing strategies and programs etc., it discusses contemporary issues pertaining to engineering consultancy. All the contemporary issues are discussed with an orientation towards decision making well supported by illustrations.

Organization

The book delineates the salient aspects of engineering consultancy and its marketing in chapters 1 to 17. The chapters have been systematically arranged and presented. An attempt has been made to agglomerate engineering consultancy marketing practices encompassing all major issues commencing with established service practices and graduating to the not–so- established practices.

Chapter 1 is historical, wherein it is established that engineering consultancy, and construction activities have been in existence since the dawn of civilization or time immemorial. It is one of the oldest *knowledge-based* professions in the world. Ever since the human race started harnessing natural resources for benefit of the mankind, the significance of this profession is being increasingly realized. Chapter 2 deals with basic understanding of engineering consultancy services, the scope of services, the advisory role an engineering consultant plays during project formulation stage as owner's design engineer and the participatory role during project execution stage as owner's site engineer, etc. Chapter 3 discusses the market paradigms, and the capricious changing buying behavior of clients. Chapter 4 describes the salient/ important features of engineering consultancy and its implication for marketing. Features like , knowledge and expertise based profession, inseparability, co-existence of service provider (Consultant) and service receiver (Client), variability /heterogeneity, intangibility, perishability, client's participation; co-operation and role contribution, low entry barriers and environmental influences have been discussed in the perspective of engineering consultancy. Chapter 5 covers ten quality elements of engineering consultancy services. Quality elements like tangibility, reliability, responsiveness, competence, courtesy, credibility, security, access, communication and understanding the client and knowledge about the project have been dealt. Chapter 6 describes the concern for quality and cost-benefit analysis of the quality of engineering consultancy services. Chapter 7 details the appointment of engineering consultants using quality and cost based selection philosophies under international competitive bidding (ICB). Chapter 8 defines engineering consultancy marketing. Chapter 9 describes the need for marketing system for engineering consultancy. Chapter 10 discusses about members of engineering consultancy marketing team. Chapter 11 analyzes internal marketing in engineering consultancy services and making it happen. Chapter 12 covers measure to be adopted for increasing marketing effectiveness. Chapter 13 covers business procurement approaches usually followed by engineering consultants for procuring jobs. Chapter 14 deals with formulating marketing mixes (8Ps) for engineering consultancy services. Chapter 15 describes an engineering consultant's contractual obligations followed by a clear conclusion in Chapter 16. Chapter 17 contains references. To make the book interesting, even to an average reader, we have made the narration terse, simple, lucid, objective, and focused. Nowhere there is a beating around the bush. Clarity in explanation is achieved by providing suitable illustrations.

Acknowledgement

At the outset, I would like to express my deep sense of gratitude to my organization MECON Limited for giving opportunities to discharge my duties and responsibilities with full freedom, leverage and liberty at various capacities in different functional areas, both at the engineering office as well as at the sites during the last 34 years of my service. The functional areas include design and engineering, inspection of plant and equipment, management advisory services, marketing and sales, consultancy; procurement engineering, contract engineering, project management consultancy (PMC), etc. This extensive and expanded engineering and project execution exposure has proved to be a source of immense knowledge, inspiration and techno-commercial professionalism for me. I owe a deep sense of obligation to my organization MECON Limited and its management. This book is a tribute to my organization MECON Limited and all the stake holders for being at this place.

During the last 34 years, I worked in close synchronization with capital goods manufacturing firms of national and international repute from across the globe, in the capacity of engineering consultant/client's engineer. The marketing, contracting, and project executives of these firms truly shared their practical wisdom during across the table discussions at pre and post contracting stages which helped in honing the subtle concepts of engineering consultancy, marketing, contracting, and project execution. I have made use of the same adequately in the book. I gratefully appreciate the insights I assimilated from them.

I also duly venerate engineers, contract personnel and project personnel of the various corporate houses/clients for their perseverance during official engineering meetings and site progress review meetings. A detailed work of this kind encompassing the myriad practices of engineering consultancy and project execution would have remained virtual without their cooperation and contribution.

Carrying out a work of this nature is an intricate task. However, I got full support from my colleagues working in various engineering departments and offices of MECON Limited. My sincere exalts are due to them for numerous discussions, both formal and informal; I had with them during the last 34 years. The fine aspects elaborated and pin pointed by them during discussions, video conferencing, kick-off meetings, project progress review meetings, at construction and erection site, etc. immensely benefited in augmenting engineering and commercial-contracting acumen. I would be failing in my duty, if I do not mention the sagacity I have gained from them.

Mere academic activity is not enough for the successful completion of a research work. It is desired that the researcher should be encouraged by the ambience around. This is what provides the strength to proceed patiently. In this respect I had received constant buttress from my peers in the Commercial Directorate and Procurement Services Section of MECON Limited. I am thankful to them for the constant buttress.

Words are not sufficient to express my heartfelt gratitude towards Sri Goutam Chatterjee, Director (Commercial), MECON Limited for forewording the book. Sri Chatterjee, in-spite of his busy schedule, spared his valuable time to go through the manuscript with all zeal. I am intellectually indebted to him for fore wording the book and grateful to him for the stanchion provided zealously.

Also, my association with executives of many organizations while engaging MDP sessions in the area of B2B industrial marketing, engineering consultancy marketing, contracting, project management, etc. has been especially rewarding. They not only provided me with an opportunity to share my conviction but also helped me sharpen my thought process. Indeed, their willingness to share ground veritable with a person plethora of times junior to them has been very efficacious. I have agglomerated their ideas in this book. I thank this group of practicing executives for their altruism in sharing their thoughts and experience.

My intellectual indebtedness is also to those academicians and practitioners who have contributed significantly to the emerging field of services marketing, Customer/ Client Relationship Marketing (CRM), Internal Marketing (IM), Key Account Management (KAM), Project Management (PM), etc. and whose ideas have been harnessed by me as lead in this book. It is my pleasure and privilege to hat tip their contribution, though it is a trivial appreciation of their contribution. I owe a deep sense of gratitude to all those who made these philosophies such an enchanting discipline.

No work worth its name can proceed without published literature of the highest grade. In this regard, I wish to express my heartfelt and ardent gratitude to the publishers of many technical, engineering and management journals for extracting the facts, figures and the opinions published therein. However the interpretations are solely mine, and I take responsibility for the divergence.

I am also grateful to my PG students of management programs for their curiosity, inquisitiveness, in-depth analysis, proactive and heretical questioning and diligence in class room lectures during the last two and half decades which proved pivotal for me to explore the unexplored realms which I had not been cognizant of.

I am indebted to my parents, other members of my family and in-laws for their inspiration, blessings and encouragement. I express my deep sense of adulation to my mother, Late Mrs. S. Chatterjee, a retired school teacher, for being the main spring of inspiration for enrichment and whose influence on my academic, professional and personal life has been profound, prolific, and indelible. I express my deep sense of gratitude to them for their constant inspiration, support, good wishes and blessings.

In my endeavor, I received tremendous academic and engineering prop from Mr. Partho Pratim Chatterjee, the co-author of this book, who converted the thought process to action by injecting and blending the hardware, the ground realities, minutiae, punctilios and nuances being a vocational trainee in MECON Limited. Mr. Partho often went beyond the scope of the book in analyzing the concepts and formulated pertinent mathematical equations to precisely quantify the concepts. The book would not have taken this fine shape without the industrious, untiring, thoughtful and dedicated effort put forth by him from engineering and management angle. This helped in value addition and right sizing of the book. Otherwise, it would have merely remained at the fluid state rather than a cast and machined product.

In writing this book, our ideas have been shaped by several persons. Each interaction enriched our experience and added a new dimension to the work. We put on record our indebtedness to all those persons who helped directly or indirectly at different points of time to carry out the work.

We also thank Ms. Mary Grace Stefanchik, Manager, Publication Development, ASME Press and Ms. Tara Collins Smith, Production Manager, ASME Press, for excellent management, whole-hearted cooperation and abet in shaping the book to its present mould and making it available to the prospective readers.

Last but not least, I express my sincere thanks to my wife Sandipa who has been a great source of strength through her affection, tender care and through the candid expression of her insights into real-life situations. Her constant inspiration, cooperation, motivation, unstinted support, bearing the tribulations, sacrifices and unflagging assistance extended ardently at every stage has helped keep going in onerous times and in sustaining the pace of the work. She deserves special acknowledgement for the same. My debt to her is prodigious and there cannot be any recompense for the same.

The words he, him, himself or man have been used in this book while referring to people in a strictly gender neutral sense. We apologize to those who find this usage objectionable. We have used it only where such usage is unavoidable or inevitable.

Further, in spite of the help and cooperation received from various quarters and despite our earnest efforts for obliteration, certain shortcomings, peccadilloes, errors, variance might have inadvertently crept in for which we humbly apologize.

Before we conclude, we would like to divulge that we would be elated to receive the invaluable opinions and suggestions on the book from the students, academia and the professional fraternity for further enrichment of the book.

Chapter 1 – Introduction

Engineering consultancy and construction activities have been in existence since the dawn of civilization. It is one of the oldest *knowledge-based* professions in the world. Ever since the human race started harnessing natural resources in the service of the mankind, the significance of this profession is being increasingly realized. The Pyramids in Egypt, the Great Wall of China, The Great Bath - Mohenjodaro and Granary- Harappa in Pakistan, the Angkor Temples in Cambodia, Tajmahal-Agra, Vijaya Vitthala Temple and Virupaksha Temple-Hampi, Brihadeshwara Temple-Thanjavur, Shore Temple and Panch Rathas-Mahabalipuram, Sun Temple-Konark, Lord Jagannath Temple- Puri, Khujaraho Temple-Khujaraho, Somnath Temple and Bet Dwarka-Saurashtra, Basilica of Bom Jesus- Goa in India, Leaning Tower of Pisa are some of the notable landmarks of the era before the industrial revolution. Wonders of the modern world like Sydney Opera House, the Channel Tunnel, Kansai International Air Port in Osaka Bay, Aswan Dam in Egypt, Sears Towers in Chicago, Kennedy Space Centre; Cape Canaveral, etc. are some of the notable landmarks of the present era.

Projects, like long span and high rise buildings like Oasia Hotel Downtown, Singapore, Chaoyang Park Plaza, Beijing, China, MahaNakhon, Bangkok, Thailand, Torre Reforma, Mexico City, Beirut Terraces, Beirut, Lebanon, Great Tower Building in Osaka City, Japan where express highway passes through the 5th, 6th and 7th floor, Kingdom Tower in the Saudi Arabian coastal city of Jeddah (1000 meter high having 160 floors whose design is both highly technological and organic with tapered wings producing an aerodynamic shape that helps to reduce structural loading due to wind vortex shedding and featuring the world's highest observation deck on the 157th floor), dams and irrigation networks, large oil refineries, integrated steel plants, super thermal power stations, long distance cross country high voltage direct current (HVDC) power transmission line, on-shore oil and gas terminals and off-shore platforms, long distance cross country oil and gas pipe lines, communication networks, satellite launching stations, industrial plants, smart cities, the proposed 1.44 km long 4 lane Silver Tunnel in London connecting North Greenwich with Royal Deck in Tames river, the 55 Km long Hong Kong - Zhuhai – Macan bridge across the Pearl River Delta in China, the 2.3 km long Pamban Railway Bridge on Bay of Bengal connecting Mandapam to Rameshwaram, Tamil Nadu in India, Train to the Clouds (Tren a Las Nubes) which is 217 kms long train service connecting Argentina's Northwest to the Chilean border on Andes Mountain Range having 29 bridges and 21 tunnels en-route in Salta province at 4.2 kms above mean sea level, 5.6 km long Bandra –Worli Sea Link in Mumbai, India, the Bhupen Hazarika Setu (9.15 km); the longest Indian river bridge spanning from Sadiya town to Dhola village in river Lohit (a tributary of river Brahmaputra) in the state of Assam; Rajiv Gandhi International Airport, Hyderabad, India, world's biggest Kaleshwaram Lift Irrigation Project, Telangana, India, etc. are few other examples of engineering consultancy and construction marvel.

Engineering consultancy services required during pre industrial revolution era were rudimentary and the projects were undertaken without time, money and resource constraints. However, these services have become highly complex requiring multidisciplinary skills in view of exacting demands with regard to high quality, economic cost and zero time over run. That is why the role of professional engineering consultant in projects is a well established fact and their appointment an essential requirement. Without the involvement of a professional engineering consultant, either in house or outside, a project can not take proper shape (Chatterjee and Sharma[1]).

Engineering consultancy services are rendered by various types of entities. These are independent individual consultants/ architects, retired technocrats/architects, research and development institutions, technical/ engineering academic institutions, in– house wing of a corporate house, corporate houses rendering consultancy/architectural services, foreign consultancy firms operating abroad, etc. These entities have experienced multi disciplinary engineering man power/architects, extensive in-house engineering data base, exposure to latest technological advancements, latest infrastructure facilities like online data

2

banks, computer hardware, engineering design and drawing software, project management software both in-house developed as well as out sourced, Memorandum of Understanding (MOU/ tie ups with leading global technology know-how suppliers, back up services from institutions like government and private research laboratories, academic institutions of repute. etc.

Chapter 2 – Engineering Consultancy Services

Engineering consultancy is basically idea imparting services. As the name suggests, it implies rendering intellect services as an expert to clients in identifying and investigating projects, imparting concepts and value added advises, recommending appropriate course of action and rendering assistance thereof to implement these recommendations. It is all about creativity, imagination and dedicated work. An engineering consultant plays an **advisory** role as *owner's design engineer* during project formulation phase and **participatory** role as *owner's site engineer* during project implementation. The scope of services of an engineering consultant may include the entire gamut of engineering activities relating to setting up of projects in green as well as brown fields. An engineering consultant is associated with the project from concept to commissioning right from project conceptualization and formulation stage to project execution and implementation stage and even during post implementation.

During project conceptualization and formulation phase, the role played by the engineering consultant is mainly *advisory* wherein the engineering consultant acts as a counselor. It persuades the client to adopt a proposal on its merit and get involved right from conceptualization stage to generation of base line data to engineering of solutions for its effective implementation. Usual services during this phase at engineering office include basic engineering, detailed engineering, procurement/contract engineering, preparation of quality assurance plan (QAP), designer's supervision, etc. These encompass any or all of the following : preparation of project profile/ concept note, carrying out market survey and arriving at product mix, preparation of pre-feasibility report/ techno-economic feasibility report (TEFR), detailed project report (DPR), process/ technology selection, site survey and site selection, raw material linkage studies, planning infrastructure facilities (mine planning, power supply, water supply, rail network, road, township etc.), preparation of environmental impact assessment (EIA)/ environmental management plan (EMP) reports, rendering assistance to client in obtaining financial, environmental and other clearances from statutory authorities, etc. It is during this stage of project that a complete dissection of all the activities of a project to the infinitesimal level is carried out and thereafter its implementation is planned in a scientific manner. The meticulous micro planning of a project done at this stage helps in execution of project without major hiccups at a future date.

During project implementation phase, the role played by the engineering consultant is mainly *participatory* wherein the engineering consultant acts as a site-counselor. It persuades the client to adopt the execution philosophy and gets involved at site right from day one for execution, construction, erection, commissioning and establishing performance guarantee of facilities. Usual services during this phase at site include organizing the project team, establishing the site office, approval of vendor's drawings, approval of billing and shipping schedules, receipt and storage of plant and equipment at site, inspection and quality assurance services at vendor's work as well as at site, project management; monitoring and control with respect to cost and time, supervision of construction, erection, testing and commissioning, invoice processing and submission of recommendation to client for release of payment to package contractor, etc. of the facilities that make projects to come up.

After commissioning of the unit/ project/plant, the consultant streamlines the operations, undertakes trouble shooting and all other activities necessary for raising plant's capacity utilization, productivity and efficiency level. Usual activities during this phase may cover any or all of the following : system stabilization, streamlining the operations, training of client's personnel in operation and maintenance (O and M), assist client in trouble shooting, coordinating preparation of 'As Built Drawings', preparation of project completion report highlighting package/facility wise history, preparation of definite cost estimate (DCE), report on problems experienced/encountered during project execution, any note worthy point, etc. and finally contract closure of packages, etc.

Chapter 3 – The Market Paradigms

Engineering consultancy firms which nurtured in a cozy environment with guaranteed business sometime back, are facing new business realities. The era of awarding contracts on nomination or cost plus basis is over. Today they are operating in an ever more complex environment with constantly changing buyer's market having stiff sales bottlenecks, wafer thin margins, from monopoly to market driven competition having presence of serious rivals, from budgetary support to self reliance, from cost plus approach to a market administered pricing, from quantity push approach to quality and timely delivery, from short term transaction to long term relationship, from life long technologies to fast changing technologies, emergence of new market players with comparable professional and technical competence, etc. Gone are the days when clients could be herded towards the services that were on offer as opposed to giving them/ providing them customized solutions what they needed.

There is a paradigm shift in client's buying behavior. Industrial business to business (B2B) clients have become more demanding, value seeking, information savvy, globalised in outlook, process and technology oriented and time conscious. Today, not only are they more demanding but are also swamped with choices. Their wide exposure to global technology and even capital have made them more demanding and discerning. They have moved resolutely from a state of low expectations to high expectations, from ignorance to full knowledge, from local access to global access, from a platform of little or even no choice to one of multiple options, from being submissive to dominating. They have strong desire to aspire for contemporary technology, plant and equipment and want value for their money/ investment. They have grown more sophisticated, demand global standard of technical and functional quality, customized offerings (8Ps), shortest project schedule, quick response to complaints/

trouble shooting, reasonable fees, payment terms based on milestones achieved, complete satisfaction of their techno- economic requirements, and so on. They have become as opportunist as can be possible and today's best may not be tomorrow's best for them. They no longer exist to have consultant's services but, rather consultants exist to serve them by solving their problems. Sometimes, they make better informed project decisions as they have wide exposure to source of information. They judge a package along bundle of benefits but not on single dimension. They have more bargaining power and are holding all the aces. They are running short of time and want more convenience. Because they assume that quality will be built into every activity, they have very little patience for poor/ inferior quality, time and cost over runs. With more choice available, they have no patience for unsubstantiated marketing mixes (8Ps) of engineering consultant. There is a sea change in their financing and procurement decisions. They stress on lesser Employer's and greater consultant's obligation. Yesterday's differential offerings has become today's essential requirement and today's best may not be tomorrow's best for them. They have become highly vocal and they are deadly allergic to aspects like furnishing inaccurate/ incomplete information both on technical and commercial matters, or exaggeration of facts/ figures in proposals, or selling on competencies not proficient in, or inability to meet commitments, or failure to settle complaints, or lackadaisical approach towards engineering/ site problems, or sacrificing quality for fees, or indecisiveness, or indifferent attitude, or communication gap, or lack of personal attention to or enthusiasm, or pre conceived ideas, or improper invoicing, or unethical practices, or holier than thou or know all attitude, or lip services, etc. They visit the design office premises of the engineering consultant prior to placement of order to assess their design, engineering and project management capabilities. Today's client want to be delighted, not just satisfied. With a plethora of alternatives available to them and attractions designed by the other competing firms to court them, engineering consultants are finding it difficult and challenging to gain, retain and regain the lost clients. The traditional concept of providing only exclusive engineering consultancy services is fast waning. As a corollary, clients are seeking services on engineering, procurement and construction management (EPCM) basis with liquidated damages (L/D) due to delay in completion of facilities and for non-fulfillment of performance guarantee (PG) parameters of the facilities. In addition to the purely technical content, there is an increased emphasis on aspects such as quality, multi-disciplinary competence and skill, execution methodology, etc.

Further, for mega projects, clients now prefer execution of the entire project on engineering, procurement, construction and commissioning (EPCC) on total turnkey basis. The thrust is to get the project executed on single point responsibility, instead of dis-aggregated or multi-spilt approach, in which consultancy, engineering, supplies, construction and erection were undertaken by different agencies. This is to ensure that overall project execution responsibility could be fixed on a single agency. Execution of projects on EPCC basis calls for mobilization of substantial capital and deployment of large human resource at engineering office as well as at site. In addition to detailed engineering and consultancy inputs, the EPCC contractor has to mobilize initial capital to take care of project's initial investment requirement namely on site preparatory work like leveling, road and drainage, construction water, construction power, civil engineering work, deployment of multi-disciplinary engineering man power starting with project engineers followed by process engineers, technologists, civil engineers, electrical engineers, mechanical engineers, instrumentation engineers, etc. Sometimes, equity infusion by the engineering consultant is also called for by the client in order to share ownership and the risk in the venture. Thus, engineering consultancy has started getting linked up with financing of the project encompassing mobilization of capital, equity participation and huge amount of working capital. Working capital of the order of 25% to 30% of the EPCC contract value is required to provide bank guarantee (BG), advance payment to vendors and to fill the gap of mismatch between cash inflow and cash outflow of project payments from client and to the EPCC contractor. In addition, the development in information technology have added 'e' flavor to every functional area of business including project management. Computer aided design engineering and drafting (CAD), computer controlled project management, process control and automation, management information system, customer/client relationship management (CRM), enterprise resource planning (ERP), SAP etc. are no longer buzz words, but have become integral part of engineering consultancy.

For mathematical understanding of market paradigms (cause and its effect), a typical example pertaining to one of the paradigms say, arrival of new state-of-the-art contemporary technology (cause), and corresponding demand fluctuation (effect) for a product like electronic control instruments is illustrated below. For illustration purpose, an attempt has been made to testify this universal hypothesis for a product like electronic control instruments. The cause and its corresponding effect can be approximately represented by equation (1) below.

As we are aware, that for a newly introduced state-of-the-art industrial product, initially the demand, being latent in nature, increases due to high acceptance rate of B2B industrial customers. The demand attains a maximum value till another product with modified vying technology comes into the domain. Due to arrival of modified vying technology, the demand of the original product starts declining and finally attains an almost constant level till it extirpates from the market.

Here, before the constant level is attained (starting from introduction phase, travelling through growth phase and passing through maturity phase of product life cycle [PLC]), the graph of the function resembles a parabola under normal circumstances. Subsequently, on attaining the constant level (at decline phase of PLC), the graph of the function approximately resembles that of a straight line curve almost parallel to X-axis with negligible slope (dy/dx → 0). The equation of the curve is derived below.

Similarly, under certain special circumstances like state- of-the- art product with iconoclastic technology and novelty to the market the graph of the function may approximately resembles that of an exponential function. For the firm or for its marketing department, it is essential to be cognizant of the maximum demand level and the year when this maximum will be achieved.

Let us assume the Equation of the parabola is

$$y = ax^2 + bx + c \tag{1}$$

In equation (1), 'y' stands for the demand level of user customers of the original/predecessor product, 'x' stands for the year from the date of introduction (date of commercial launching) of the original/ predecessor product (i.e from zero date). The values of a, b and c are constant and can be arrived at using the standard curve fitting equations (1a), (1b) and (1c) below.

$$\sum y = nc + b\sum x + a\sum x^2 \tag{1a}$$

$$\sum xy = c\sum x + b\sum x^2 + a\sum x^3 \tag{1b}$$

$$\sum x^2 y = c\sum x^2 + b\sum x^3 + a\sum x^4 \tag{1c}$$

In equation (1a), *'n'* stands for number of data points collected from market research for the concerned variable. More the data points collected, better the accuracy of the curve.

The numerical values of *a*, *b* and *c* thus obtained from the above three equations (1a), (1b) and (1c), can be substituted in equation (1) and the fitted parabola can be plotted.

To precisely know the year in which the demand level will be maximum, we have to first differentiate *y* in Equation (1) w.r.t *x* (*i.e. dy/dx*) and then find *x* when *dy/dx = 0*

$$y = ax^2 + bx + c$$

$$\text{or } dy/dx = 2ax + b$$

Now, equating *dy/dx = 0*, we get,

$$2ax + b = 0,$$

$$\text{or } x = -b/2a$$

Now, differentiating *y* w.r.t *t* again, we get,

$$d^2y/dx^2 = 2a$$

For *y* to have a maxima at $x = -b/2a$, d^2y/dx^2 should be negative. One can check, if d^2y/dx^2 is negative for $x = -b/2a$. This clearly implies that the function (demand level) has a maxima at that value of $x = -b/2a$ (which means $a < 0$). Thus, the time **x** when the product demand level will be maximum is obtained (*i.e.*, $x = -b/2a$). Now, substituting this value of time **x** in the equation of the parabola $y = ax^2 + bx + c$, maximum demand level *y* can be arrived at. As the curve passes through origin, the value of *c* is 0.

As described above, on attaining the constant level (at decline phase of PLC), the graph of the function approximately resembles that of a curve almost parallel to X-axis with negligible slope (*dy/dx* → 0). The equation of such a curve with negligible slope is derived below. We have:

$$dy/dx = (1\text{-Log } x)/x^2$$

$$\text{Or, } dy/dx = (1/x^2)\text{- (Log } x)/x^2$$

$$\text{Or, } \int dy = \int [(1/x^2)\text{- (Log } x)/x^2] \, dx$$

$$\text{Or, } \int dy = \int (1/x^2)dx\text{- (Log } x)/x^2 \, dx$$

$$\text{Or, } \int dy = \int (1/x^2)dx\text{- (Log } x)/x^2 \, dx$$

$$\text{Or, } \int dy = \int (1/x^2)dx - (\text{Log } x)/x^2 \, dx$$

$$\text{Or, } y = (-1/x) - \int (\text{Log } x)/x^2 \, dx + c$$

Using integration by parts, we have,

$$y = (-1/x) - \int (\text{Log } x)/x^2 \, dx + c$$

$$\text{Or, } y = (-1/x) - \int (\text{Log } x) \, x^{-2}dx + c$$

$$\text{Or, } y = (-1/x) - [(\text{Log } x) \int x^{-2}dx] - \int\{d(\text{Log } x)/dx \, {}^*\int x^{-2}dx\}dx] + c$$

$$\text{Or, } y = (-1/x) - [-(\text{Log } x) \, x^{-1}] - \int\{-1/x \, {}^* x^{-1}\}dx] + c$$

$$\text{Or, } y = (-1/x) - [-(\text{Log } x) \, x^{-1}] - \int\{-1/x \, {}^* x^{-1}\}dx] + c$$

$$\text{Or, } y = (-1/x) - [-(\text{Log } x) \, x^{-1}] - \{x^{-1}\}] + c$$

$$\text{Or, } y = (-1/x) + (\text{Log } x) \, x^{-1} + x^{-1} + c$$

$$\text{Or, } y = (-1/x) + (\text{Log } x) \, x^{-1} + (1/x) + c$$

$$\text{Or, } y = (\text{Log } x)/x + c$$

Where, c is a positive constant.

The domain of the function in years may be, say (0, 4] or (0, 5] or (0, 6], (0,7], (0, 8] or (0, 10] depending upon product and its functional capabilities, market's ability to pay, market's willingness to pay, market's technology level (developed country vs developing country) capacity of the market, market's exposure to the source of information etc.

Similarly, under certain special circumstances, like state- of- the- art product with iconoclastic technology and novelty to the market, the graph of the function may approximately resemble that of an exponential function. For such product, having exponential increase in demand with time the curve of demand verse time approximately represents an exponential curve instead of a parabola (during introduction and growth phase of PLC). In such eccentric cases, slope of the curve is represented as,

$$dy/dx - xy = x$$

This first order linear differential equation can be solved using Integrating Factor (IF).

$$\text{IF} = e^{\int -x \, dx}$$

$$\text{Or, IF} = e^{-x^2/2}$$

The solution of such differential equation can be given by:

$$y * IF = \int (x * IF) \, dx + c$$

$$\text{Or, } y \, e^{-x^2/2} = \int (x * e^{-x^2/2}) \, dx + c$$

$$\text{Or, } y \, e^{-x^2/2} = \int (x * e^{-x^2/2}) \, dx + c$$

$$\text{Let } -x^2/2 = z, \text{ we have,}$$

$$-x \, dx = dz$$

Using integration by substitution,

$$y \, e^{-x^2/2} = \int (-e^z) \, dz + c$$

$$y \, e^{-x^2/2} = \int (-e^z) \, dz + c$$

$$y \, e^{-x^2/2} = -e^z + c$$

$$y \, e^{-x^2/2} = -e^{-x^2/2} + c$$

$$y = -1 + c \, e^{x^2/2}$$

As the curve passes through origin, the value of c is 1.

Insight/inference can be drawn about the future course of action, that when (i.e time x) the demand level **(y)** will be at peak/ zenith. The firm can, accordingly formulate its business plan and strategy in various functional areas for capacity enhancement and subsequently for introduction of modified brand or new model i.e. brand extension or new product introduction accordingly.

If an entity has to do well in the cut throat engineering consultancy segment, it must address on the paradigms and provide solutions that are relevant, cost effective and efficient.

Chapter 4 – Features of Engineering Consultancy Services and Its Implications for Marketing

Engineering consultancy has a number of unique features which makes it different from industrial products or capital goods. Taking a lead from the writings of authors [2–4], following unique features of engineering consultancy services are being proposed. The features are Knowledge and expertise based profession, Inseparability, Co-existence of service provider (consultant) and service receiver (client), Variability/heterogeneity, Intangibility, Perishability, Client's participation; co-operation and role contribution, Low entry barriers, Environmental influences, etc. Each of these features has a number of managerial and operational implications which makes its marketing idiosyncratic with respect to industrial products or capital goods. It is the combination of these features that create the specific context in which an engineering consultancy firm formulates marketing policies and strategies. A clear understanding of the features is an essential pre-requisite for formulating client centric marketing strategy.

4.1   Knowledge and Expertise Based Profession

Engineering consultancy is multi-disciplinary knowledge and expertise based profession. An engineering consultant is one who possesses the required knowledge, analytical power, expertise and skill, one who renders advisory services or participatory services or both to client at different stages of project execution, one who is fully conversant with the magnitude of the engineering and technological issues likely to be encountered in execution of the project, one whose opinion can be relied upon when ideas are conflicted or bounced against each other. The personnel of engineering consultant are expected to play the role of a partner, teacher, advisor, reviewer, facilitator, motivator, negotiator (with input suppliers or vendors) to give shape to the ideas into engineering realities in line with the best industry practice. To this end, the client seeks the credentials of the engineering consultants and asks them to furnish list of similar assignments/project executed in the recent past for other clients.

4.2   Inseparability

Engineering consultancy services are inseparable as the engineering consultant and the services he provides cannot be separated and the services are consumed as it is produced. This also means that the service provider (consultant's engineer) becomes a part of the product/service itself and finds himself as an essential ingredient in the service experience for the client.

4.3   Co-existence of Service Provider (Consultant) and Service Receiver (Client)

In engineering consultancy, co-existence of service provider (consultant) and service receiver (client) is a very common phenomenon. Figure 1 depicts the co-existence of engineering consultant and client in project. As depicted, on most of the occasions, the consultant/its project team and client/its project team work in close association with each other during different phases of project execution at various places. It may be at design and engineering office of the consultant, or at the office of the statutory authorities for defending the project report, or at works/shop floor of equipment manufacturer or at project site of the client etc. During project conceptualization, project appraisal, project planning and project execution stage, consultant's engineering team and client's project team operate in closely knit manner and are involved in simultaneous service transaction. Engineers of engineering consultant interact with their counter parts on a

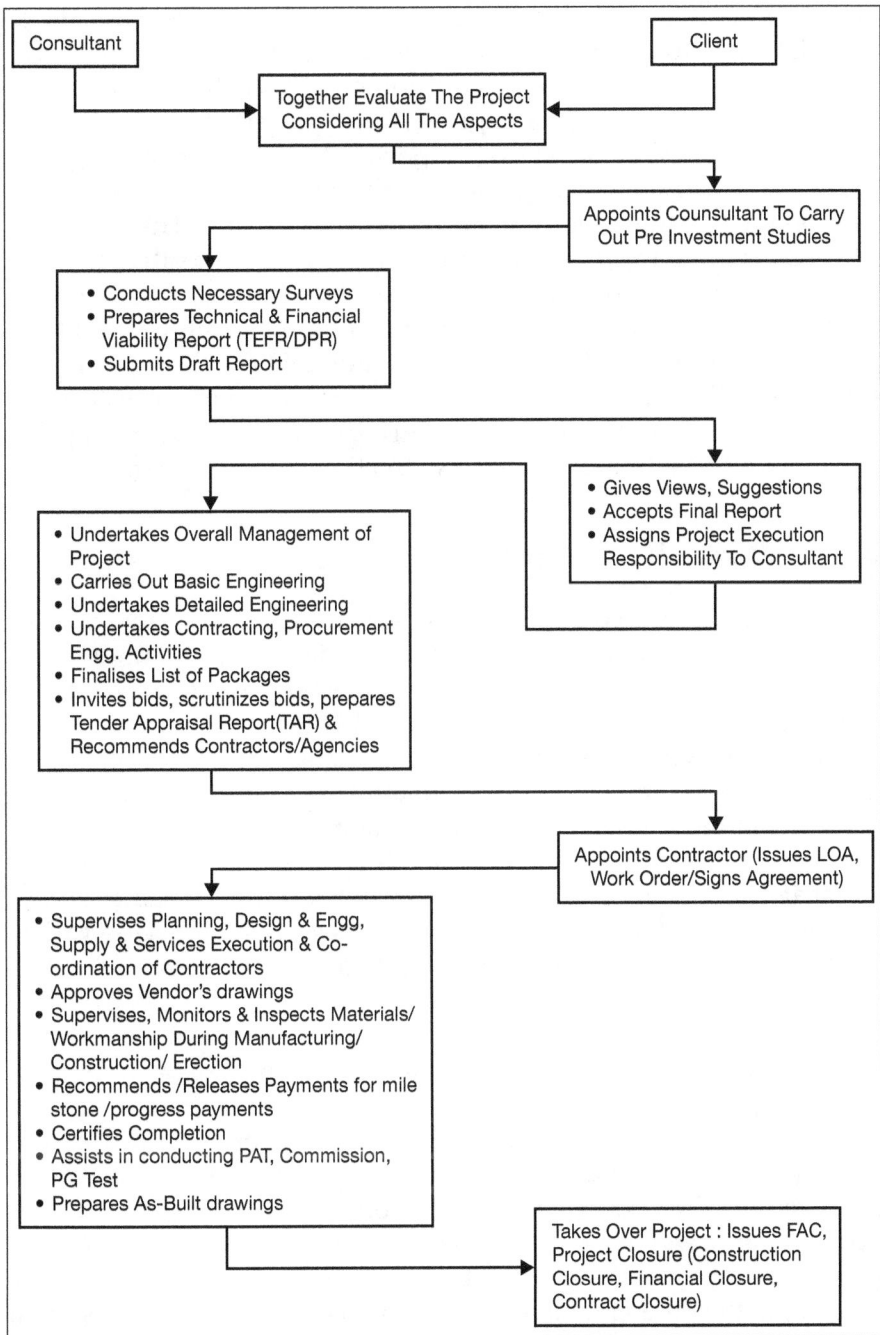

```
┌──────────────┐                                          ┌──────────────┐
│  Consultant  │                                          │    Client    │
└──────┬───────┘                                          └──────┬───────┘
       │        ┌────────────────────────────────┐               │
       └───────→│ Together Evaluate The Project   │←──────────────┘
                │ Considering All The Aspects     │
                └────────────────────────────────┘
```

Together Evaluate The Project Considering All The Aspects

Appoints Counsultant To Carry Out Pre Investment Studies

- Conducts Necessary Surveys
- Prepares Technical & Financial Viability Report (TEFR/DPR)
- Submits Draft Report

- Undertakes Overall Management of Project
- Carries Out Basic Engineering
- Undertakes Detailed Engineering
- Undertakes Contracting, Procurement Engg. Activities
- Finalises List of Packages
- Invites bids, scrutinizes bids, prepares Tender Appraisal Report(TAR) & Recommends Contractors/Agencies

- Gives Views, Suggestions
- Accepts Final Report
- Assigns Project Execution Responsibility To Consultant

Appoints Contractor (Issues LOA, Work Order/Signs Agreement)

- Supervises Planning, Design & Engg, Supply & Services Execution & Co-ordination of Contractors
- Approves Vendor's drawings
- Supervises, Monitors & Inspects Materials/ Workmanship During Manufacturing/ Construction/ Erection
- Recommends /Releases Payments for mile stone /progress payments
- Certifies Completion
- Assists in conducting PAT, Commission, PG Test
- Prepares As-Built drawings

Takes Over Project : Issues FAC, Project Closure (Construction Closure, Financial Closure, Contract Closure)

Figure 1 – Co-existence of Engineering Consultant and Client in Project.

number of occasions for various engineering and non engineering issues. Such interactions are more during project execution phase as compared to design and engineering phase. And, each such individual interaction creates a Moment of Truth [5] about the engineering consultant, its engineers, its quality of work, etc. and provides/withholds marketing opportunities. Thus, each engineer involved in service transaction has the opportunity to favorably reinforce or change client's perception and beliefs. Their attitude towards client's project team at every interaction creates bonding (or un-bonding). Any variance (positive or negative) by any one of them at any instance has the potential to either gain client's trust or have the opposite effect. It is the moment where the consultant's image is built or destroyed. A series of moments of truth leads to client delight or annoyance. Engineering consultant must keep this aspect in mind while interacting with client's project team.

4.4   Variability / Heterogeneity

Engineering consultancy services are highly variable as such services depend upon the engineer who renders them and under what circumstances they are being rendered. The services are personalized in nature as human element is associated with it. Due to this reason, quality of service delivery keeps on changing from engineer to engineer, site to site, time to time, situation to situation, and the service (as experienced) may not be of a consistent quality. What is excellent for one client, may not be so for another, or what is excellent for one client at a particular occasion/time, may not be so for the same client at another occasion/time. The profession represents such a wide variety of expertise that it is impossible to put them into any common system of grading.

Due to variability/heterogeneity, engineering consultants face a major problem, which is to maintain consistent service quality. Uniform performance standards are difficult to set. Service quality can't be guaranteed despite the fact that guidelines and procedures have been laid down to carry out design and engineering activities, preparation of assignments, preparation of technical specifications, approval of drawings, etc. to minimize the effect of human element. Clients are aware of this variability aspect and hence before selecting a consultant they prefer taking views of others who have availed similar services from the engineering consultant.

4.5   Intangibility

Engineering consultancy services are intangible in nature. As such the client does not get anything tangible. Being intangible, it can't be displayed, seen, tasted, felt or experienced before they are consumed/availed/rendered. It can only be experienced from the effect/benefits created on client's operation/business system. In case of capital goods/industrial installation, the value can be assessed at pre-purchase period but in case of engineering consultancy, the value of consultants' services cannot be assessed at pre-purchase period but can be assessed only on its completion/consumption at post-purchase period. It is difficult to provide an exact sample of the engineering consultancy services that is on offer. A new client's experience of engineering consultancy service is in totality, the effect of all the elements present at the time of rendering service. It is not possible to duplicate the totality as a sample even the entire identical infrastructure and resources (engineers) are deployed and put in place. As such quality of the service varies and marketing mixes are difficult to formulate.

Further, like any other services, acquiring engineering consultancy services is associated with high degree of trust due to high degree of perceived risk. To decrease risks, to increase client's confidence level and to make services differentiable with respect to competitors, engineering consultancy firm must increase tangibility indirectly. Deployment of right set of engineers for the project, physical evidences like full-fledged engineering office with all latest infrastructure, IT tools, Auto CAD, plant engineering and design software, project management software, piping flexibility and stress analysis software, equipment design software, electrical system study software (viz. ETAP, AutoGridPro, PSCAD, PLSCAD), etc. play a vital role in increasing/adding the tangibility.

4.6   Perishability

Because of simultaneous production and consumption, engineering consultancy services like site-supervision services are perishable, they cannot be stocked for later consumption and hence non-inventorial. They tend to go waste/unutilized, if not availed during contractual time schedule. The loss of money on account of say, non utilization of consultant's engineering manpower deputed at site for rendering designer's supervision services for construction, erection, preliminary acceptance test (PAT), commissioning, performance guarantee (PG) tests, etc. during contractual time schedule cannot be made up.

4.7 Client's Participation, Co-operation and Role Contribution

Client's participation, co-operation, involvement and role contribution is important for successful discharging of responsibilities and meeting contractual responsibilities/obligations by the engineering consultant. It is the key to the quality of service and project execution success. Quality of service depends to a great extent on the client's attitude towards consultant's project team, disclosure of their requirement, furnishing of techno commercial information, site's geo-technical data, taking prompt decisions etc. For the engineering consultant, client's project team is his extended arm.

In fact, without client's involvement, inputs, co-operation and prompt action, the quality of engineering consultancy service may be far from usurp. Due thought in project formulation, disclosure of relevant information, furnishing of complete field data in time, adequate efforts, full involvement of project department of client, no or minimal midstream changes/modifications in project parameters, or insisting to changes which are inevitable, managing clearances from the various statutory/regulatory authorities, quick action after project clearance, making the front available and giving "Go-Ahead" clearance, identification and empowerment of the key project personnel, mobilization of fund, no change in project team, avoidance of multiplicity of agencies/vendors as far as possible, placement of trained manpower and input materials/feedstock in position before PAT, commissioning and conducting of PG test, transparency of thoughts and working, etc. are important actions/obligations which a client has to take/fulfill for successful and timely completion of the project.

Likewise, engineering consultants are equally important for the client. Their role being owner's engineer, they should be considered as integral part of their project team. To be effective, they have to be treated as an insider and shall be given full co-operation, power and authority. The project will succeed if service composition from client's perspective and consultant's perspective are in consonance. The consultant could only bring about the change if the client is willing to listen, especially to unpalatable truths. To bridge this gap engineering consultants should try to assess what benefits the client seeks, their expectations from the consultant as well as package contractor, etc. during pre engineering stage while finalizing the scheme and rendering them well during contract execution stage and even after project commissioning and stabilization stage. The challenge for engineering consultant is to solicit co-operation from the client's project team.

4.8   Low Entry Barriers

Engineering consultancy services need moderate investment. The infrastructure required is not capital intensive. As a result, the barriers to entry are low rendering the industry assailable/vulnerable/exposed to stiff/nasty/wafer thin competition, both in terms of number and magnitude. One can find independent individual consultant/freelancers in left hand side (LHS) of the fulcrum to large corporates on right hand side (RHS) of the fulcrum. The biggest challenge for a large engineering consultancy firm who is at the RHS of the fulcrum is mushrooming of consultants with lower fixed costs and outsourcing models, which enables them to bid aggressively for the limited opportunities.

Low entry barriers (cause) influence nature, rate and degree of competition (effect). This phenomenon of *Cause and Effect* can be understood quantitatively using mathematical equation. The cause and its corresponding effect can be approximately represented by equation (1) below. The equation is analogous to **Arrhenius equation in Chemical Kinetics** which describes the effect of activation energy and temperature (causes) on reaction kinetics (effect). For illustration purpose, an attempt has been to testify the universal hypothesis that when barrier decreases and business friendliness parameter increases, the number of competitors increases.

$$K = Ae^{(-Q/RT)} \tag{1}$$

Where,

K = Number of competitors

Q = Entry barriers

T = Business friendliness parameter, which depends on external business environmental factors like say, technology awareness level of market/customer, Government's industrial and foreign trade policy, political stability, economic condition of market, law and order situations, etc.

A = Numerical Constant

e = Euler's number = 2.7182

R = Universal Constant = 8.314

Now, differentiating K w. r. t. Q, (i.e. rate of change in number of competitors due to intensity change in barriers), we get,

$$dK/dQ = Ae^{(-Q/RT)} \times (-1/RT) \qquad (2)$$

As the RHS in equation (1) contains e, we take Natural Logarithms (instead of Common Logarithm) on both sides for simplification. We thus obtain,

$$Ln\ K = Ln\ A - Q/RT \qquad (3)$$

$$Or,\ Ln\ K_1 = Ln\ A - Q_1/RT, \qquad (4)$$

Situation (A) when barriers are higher

$$Or,\ Ln\ K_2 = Ln\ A - Q_2/RT, \qquad (5)$$

Situation (B) when barriers are relatively moderate (Q2 < Q1)

Subtracting equation (4) from equation (5), we get,

$$Ln\ K_2 - Ln\ K_1 = (Q_1 - Q_2)/RT \qquad (6)$$

$$Or,\ Ln\ (K_2/K_1) = (Q_1 - Q_2)/RT$$

$$Or,\ K_2 = K_1\ e^{(Q_1 - Q_2)/RT} \qquad (7)$$

From equation (7), it is evident that when barriers decrease, [i.e from situation (A) to situation (B)], $(Q_1-Q_2)/RT$ is positive. This indicates that theoretically the number of competitors increase as K_2 works out to be greater than K_1. Hence, the hypothesis that with decrease in barriers, the number of competitors increases, is testified mathematically.

Further, differentiating K w. r. t. T, (i.e. rate of change in number of competitors due to business friendliness), we get,

$$K = Ae^{(-Q/RT)} \qquad (1)$$

22

Or, $dK/dT = [A \times \{d(e^{(-Q/RT)} /d(-Q/RT)\}] \times [d (-Q/RT)/dT]$

Or, $dK/dT = [Ae^{(-Q/RT)}] \times [(Q/RT^2)]$

Or, $dK/dT = Ae^{(-Q/RT)} \times (Q/RT^2)$

Or, $dK/dT = K \times (Q/RT^2)$, as $K = Ae^{(-Q/RT)}$

Solving the differential equation,

$$\int dK/K = \int (Q/RT^2) \, dT$$

Or, $[Ln \, K] = [-Q/RT]$ where K varies from K_1 to K_2 and T varies from T_1 to T_2

$$Or, \, Ln \, K_2 - Ln \, K_1 = Q/R [(1/T_1) - (1/T_2)]$$

$$Or, \, Ln \, (K_2/K_1) = Q/R [(1/T_1) - (1/T_2)]$$

$$Or, \, K_2 = K_1 \, e^{[Q/R \{(1/T_1) - (1/T_2)\}]} \tag{8}$$

As business friendliness parameter increases, $T_2 > T_1$. Therefore, clearly, $Q/R(1/T_1-1/T_2)$ is positive and hence $K_2 > K_1$. So, the hypothesis that with increase in business friendliness, the number of competitors increases is testified mathematically.

Conclusion: Equation (7) and equation (8) testifies the hypothesis, that when barriers decrease and business friendliness parameters increase, the number of competitors increases, provided other market conditions/operating parameters remains unaltered.

4.9   Environmental Influences

As in any other industry, the engineering consultancy service industry too is substantially influenced by the environmental development such as alteration in Government's industrial policy, export-import trade policy, World Trade Organization (WTO) norms, arrival of new technology, etc. Such changes lead to change in the industry structure, size of service firm and the intensity of competition. Regulation of Government policies with regard to mergers and acquisition are leading to changes in the scope and scale of operation in many engineering consultancy service firms, implying different levels of competition and range of services offered.

Chapter 5 –
Quality Elements
of Engineering
Consultancy Services

Defining quality of services has never been easy. For someone, quality refers to certain standards, ways and means by which those standards are achieved, maintained and improved upon. For others, it is the totality of features and characteristics of a product or services that bears on its ability to satisfy stated and implied needs. For some one other, quality means fitness for use, fitness for purpose, fitness for function, etc. For a marketing manager, quality means customer coming back with appreciation and placing repeat orders rather than product coming back with complaints / castigations.

Gronross [6] proposed two types of service quality: technical quality and functional quality. Technical quality involves what customers actually receive (i.e. what is delivered or outcome of the service) whereas functional quality refers to the manner in which the services are rendered to the client (i.e. how it is delivered or the process of service delivery).

In engineering consultancy, the technical quality may encompass issues like conformance to technical parameters of design, engineering, construction, erection and commissioning as per client's project requirements/ personalized expectations, a user friendly tailor made project management system providing necessary flexibility of operation, a technology that is proven, prevailing industry practices, dedicated and competent human resource, etc. Whereas functional quality refers to the manner in which the engineering consultancy services are rendered to the client. The functional quality may encompass aspects like responsiveness, ability to render services reliably, ability to meet deadlines, cost effectiveness, communication, knowledge sharing etc. Majority of engineering consultancy assignments come by word of mouth based on consultant's past performance in carrying out similar assignments in quality way.

To have insight about the quality of service, Zeithaml, Parasuraman and Berry [7] conducted a series of consumer focus group interviews. From focus group research, they identified ten general dimensions that represent the evaluative criteria customers use to assess service quality. The ten quality evaluative criteria are tangibility, reliability, responsiveness, competence, courtesy, credibility, security, access, communication and understanding the customer/ client. In subsequent research, they found a high degree of co-relation between many of these variables. They consolidated them into five broad dimensions namely tangibles (appearance of physical elements), reliability (dependable, accurate performance), responsiveness (promptness and helpfulness), assurance (competence, courtesy, credibility and security) and empathy (easy access, good communication and customer understanding). To measure customer elation with different aspects of service quality, Parasuraman, Zeithaml, and Berry [8] developed a survey research instrument called SERVQUAL. It is based on the premise that customers can evaluate firm's service quality by comparing their perceptions of its services with their expectations. Later, SERVQUAL became the most commonly cited generic measurement tool that can be applied across a broad spectrum of service industries for measuring service quality. In its basic form, the scale contains 21 perception items and a series of expectation items, reflecting the five dimensions of service quality described above.

Later, SERVQUAL has been modified to E-QUAL in e-business giving opportunity to consumers to evaluate the firm's service quality performance on line. The seven dimensions of E-QUAL are Accessibility (ease to access the site), Navigation (ease to move around the site), Design and presentation (colors, layout, originality), Content and purpose, Responsiveness (firm's promptness to respond to e-mails), Interactivity, customization and personalization (relates to empathy dimension of service quality), and reputation and security (customer confidence issues) [9].

For engineering consultancy services, quality is not uni-dimensional. It encompasses numerous factors. In line with the philosophies proposed by Zeithaml, Parasuraman and Berry, the quality of engineering consultancy services can also be broken down into ten dimensions which can be used to evaluate the quality. The ten dimensions of service quality pertaining to technical and functional aspects which can be adapted to evaluate the quality of engineering consultancy services are tangibility, reliability, responsiveness, competence, courtesy, credibility, security, access, communication and understanding the client and knowledge about the project. An engineering consultant's ability to hang on to its client depends on how consistently it delivers service to them on these aspects.

5.1 Tangibility

Tangibility refers to the physical evidence and infrastructure facilities of the engineering consultant. It includes consultant's resources, full-fledged design and engineering office with latest infrastructure, IT tools, computer hardware and software, AutoCAD, plant engineering and design software, project management software, piping flexibility and stress analysis software, equipment design software, electrical system study software (viz. ETAP, AutoGridPro, PSCAD, PLSCAD), site office set up, technical information centre, testing tools, etc. to be used to render the services. As engineering consultancy services are intangible, the incorporation of tangibles either directly or indirectly creates an impression to the client about the standard of service they may get from the engineering consultant. For example, a design and engineering office in a shabby building with out-dated computer hardware and software, lack of standardization, a small technical information centre etc. will make the client wonder whether the engineering consultant will be in a position to execute the project in the right manner and within contractual time schedule.

5.2 Reliability

Reliability refers to the ability of the engineering consultant to perform the committed service dependably and accurately. It includes consistency of performance and dependability on engineering consultant's advice. Accurate design free of errors, no or minimum mid stream modifications/ changes, modifications/ changes which are inevitable due to site conditions, correct approval of vendor's drawings, proper record keeping, performing the service at the designated time, proper site supervision, bringing the variances into the notice, correct progress/quantity verification, suitable remedial measures, correctness of 'As Built Drawings', accuracy in work certification, accuracy in bill/invoice recommendation, etc. are the reliability parameters. It also includes keeping time and cost commitment and honoring the promises. Consistency of performance comes from the control of processes that delivers the service at levels that are predetermined.

5.3   Responsiveness

Responsiveness concerns engineering consultant's willingness, promptness, and readiness to render services or attending to client's request promptly. It involves timeliness of service, being proactive rather than being reactive, reverting back quickly, deputation of right mix of engineers to the project site at the earliest, attending to specific site issues, giving prompt justification/ solutions, etc. Engineering consultancy service is a function of consultant's behavior, attitude, empathy and response. And, a responsive person is one who is willing to the attend to the client's request happily and has client's project interest in his heart. He is prompt, pro-active and ready to go-ahead and does not find the first available excuse to say 'No' or 'If' or 'But'. Responsiveness comes from the concern for client's property/ investment, infrastructure and resources and a genuine interest in successful commissioning of the project. No amount of technology can compensate for dearth of responsiveness towards quality of service.

5.4   Competence

Competence with respect to engineering consultancy means possession of required know–how, skilled and experienced multi-disciplinary engineering man power, dedicated project group, efficient site construction, supervision and project management consultancy (PMC) team, etc. to perform the services. Successful consultancy implies rich engineering, technology and project experience and high all-round capabilities with confidence. For example, if the site engineer is observed to be inadequately informed about the site associated engineering issues and is seen to be wavering, repeatedly referring to the Resident Engineer or to the engineering head office, or trying to hide his ignorance, there has to be strong dissonance. This amounts to lack of consultant's concern for the project. It is true that it is not possible for every engineer to have customized/ tailor-made solutions for every engineering issue. But if the concerned engineer, cognizant of his ignorance on a particular issue, is quick to obtain the correct information from the engineering head office, and takes appropriate decision deemed fit, he is considered to be competent.

5.5 Courtesy

The politeness, consideration, friendliness, caring, feeling, affability, co-operative attitude and respect shown by the engineering consultant's personnel to each and every member of client's project team during various occasions can be bundled into the term courtesy. It is what the client perceives at a particular time, not before or after. One does not have to be servile while being courteous. While disagreeing with the client's project team on an engineering issue, it is not necessary to declare that 'I do not agree' or ' You are wrong' or 'It is just not possible' etc. It is possible to be quite assertive while stating a point of difference and yet be courteous.

5.6 Credibility

Credibility refers to the perceived trustworthiness, right conduct, honesty, sincerity, faithfulness, confidence, etc. of the consultant's engineers. Contributing to credibility are meeting contractual obligations, past reputation, quality of interaction with the client, etc. Reputation is what the consultant has to en-cash. Majority of important engineering consultancy assignments come by word of mouth based on the consultant's past credential and performance.

5.7 Security

Security refers to the freedom from danger, damage, risk, doubts, malfunctioning, etc that a client seeks from the engineering consultant. It includes confidentiality of drawings and documents, proprietary items as well as operational safety of the process, technology, no extra work, no rework, avoidance of risk and purchase situations, etc.

5.8 Access

Access refers to the approachability and ease with which the client can contact the engineering consultant. Convenient location of engineering office, opening of site office, matching working hours,

24 x 7 hours site working, short waiting time, e-connectivity, etc. are the accessibility parameters. Proximity is not determined by the physical distances between the project site and the engineering office of the consultant as Information Technology (IT) revolution has cut down distances and time.

5.9 Communication

Communication refers to keeping client informed on matters relevant to the project execution in the manner they can understand and subsequently listening to them. This includes accurate and timely information, transparency in approach, use of the correct language which is neither rosy nor harsh, etc. Increased level of sophistication with a well established knowledgeable client could be a good communication strategy. Submission of correct proposal, furnishing clarifications to client's queries, daily /weekly/ monthly project progress review reports, non compliance report, variance report in the desired format, etc. to client in time is perhaps the best way to communicate.

5.10 Understanding the Client and Knowledge about the Project

It refers to consultant's knowledge about the client, its project team, its decision making units, their expectation from the project investment, individual members' expectation, their sentiments; feelings; constraints; compulsions, etc. and clarity about the project execution methodology to avoid mid terms changes. A clear understanding on all these aspects is vital for successful completion of the assignment. As mentioned earlier, client's active participation and involvement is the key to the quality service and to solicit client's co-operation, a clear understanding about all these aspects is a must. In fact, without clear understanding, the service quality may be far from appropriate.

The quality of engineering consultancy services is thus situation specific and largely defined by client. It begins and ends with client and its project team. Client's project team is the final arbiter of quality as they evaluate service not only on its outcome but also the process associated with it.

Chapter 6 – Concern for Quality: Cost-Benefit Analysis of Quality of Engineering Consultancy Services

The concern for quality service for engineering consultancy has risen to unprecedented levels. In engineering consultancy, price/fees is forgotten very quickly but quality is never forgotten. In engineering consultancy, if one thinks good design and engineering is expensive, he should first look at the cost of poor design and engineering. Evaluating the cost of poor service quality is important as it helps in recognizing the value of maintaining quality. Poor quality impacts overall cost, as cost of poor quality has tangible and non tangible ramifications and hence substantial.

In engineering consultancy, cost of poor quality incurs at two levels, namely within the firm and at the client's end. Cost incurred at firm's end are experienced before the services are delivered to the client. These costs include cost of correcting defects like design modification identified before the drawings/ documents/ reports are issued, de-motivation of design engineers, deterioration of work culture in engineering office due to frequent failure in quality, etc. Cost incurred at the client's end are experienced after the services are delivered. These include those costs that are incurred when a defect is identified by client and brought to the notice of the engineering consultant resulting in incurring expenditure like dismantling, re-construction and re-erection work, replacement of component, redeployment of resources, cost of time over run, risk and purchase notice (RPN), inconvenience caused, penalty charges, de-motivation of site engineers, reduced productivity and dampening of work spirit at site, client dissatisfaction and annoyance, cost overrun of client's money due to extension of project gestation period, loss of revenue earning/ net sales realization (NSR) at client's end due to delay in commissioning and hence commercial production, etc.

In engineering consultancy, improving quality of the services should be viewed as an ongoing process, rather than as an occasional activity or intermittent activity or hiccupping affairs etc. The cost of maintaining quality is not high. The cost elements for maintaining quality comprises of appraisal costs and prevention costs. Appraisal costs include cost of checking, inspection and auditing of design, engineering and site supervision procedures. Prevention costs include costs incurred to reduce or eliminate the chances of poor quality and cost of employing the right or qualified people, cost of training and retraining, cost of supervision and last but not least, the amount client will recovery from consultant's fees as liquidated damages due to delay in completion of facilities and performance.

The incorporation of quality elements in engineering consultancy services is an important factor for client delight. One of the major ways an engineering consultancy firm can differentiate itself is by delivering consistently high quality service than its competitors. As explained earlier, superior quality leads to reduction in overall project cost and consequent client delight which in turn affects future buying intentions and hence business. Quality service can give a potent patronage and hence competitive advantage that may lead to placement of repeat order by the client to the consultant on nomination/ single party basis.

Chapter 7 – Appointment of Engineering Consultant Under International Competitive Bidding (ICB)

It is universal fact that without the assistance of consultant, either in-house or external, a project cannot take shape. Some projects which are routine or repetitive in nature for which design and engineering is already available, services of external engineering consultant may not be required, or even if it is required, it may be restricted to seeking second opinion or designer's supervision. The need of an external engineering consultant arises when there is no in-house expertise available in the firm or when the in-house expertise is not sufficient to meet the requirement of the project or as the investment in the project is humongous; so is the gestation period and so is the perceived risk, or that a unique technology is involved in the project, or that multi disciplinary engineering is involved, or that the project is executed on the basis of imported technology and know-how, or that multiple agencies/sub-contractors/vendors are involved, or a brown field project, etc.

Engineering consultants are appointed through international competitive bidding (ICB), national competitive bidding (NCB) process wherein client issues request for qualification (RFQ) application, notice inviting tender (NIT)/ request for proposal (RFP), and seeks proposal as per terms of reference (TOR) stipulated in NIT/RFP. Engineering consultant are required to submit their proposal comprising Part-1 (Techno-commercial Proposal) and Part-2 (Financial Proposal: Price/ Fees) separately. Initially, Part -1 (Techno-commercial proposal) of the bidders are opened and technical and financial pre-qualification (PQ)/ eligibility criteria prescribed in the RFQ/NIT/RFP are evaluated first.

The technical PQ/eligibility criteria of the bidder/engineering consultant usually includes either a few or a combination, or all of the following depending upon the nature of the assignment: **(i)** experience of rendering engineering consultancy services encompassing similar scope (defined in the NIT) for at least one number of similar project/ assignment as mentioned in NIT/RFP during last 7 years ending previous day of last date of submission of proposal **(ii)** the project/ assignment should have been commissioned/completed prior to at least one year ending previous day of last date of submission of proposal **(iii)** Should have been in the business of rendering consultancy services thereafter reckoned from the date of qualifying order as referred above, etc. etc.

The financial PQ/eligibility criteria of the bidder/engineering consultant usually includes either a few or a combination or all of the following depending upon the nature of the assignment : **(i)** Annual turnover i.e. average annual gross turnover from engineering consultancy services of last three consecutive financial accounting years ending the immediate last financial year in which RFP/NIT is invited shall be US\$ X (arrived at on certain percentage of the estimated cost of the package on annualized basis), **(ii)** Solvency i.e. Solvent for a value not less than US\$ Y [(arrived at on the estimated cost of the package divided by projects duration in month) × 3 months (say)], **(iii)** Net worth i.e. Positive net worth in each of the three previous financial years, **(iv)** Working capital, which shall be at least 't' times the monthly cash flow requirement and shall not be less than US\$ Z [(arrived at on the estimated cost of the package divided by projects duration in month) × t times], **(v)** Profit, i.e. Earned profit, and not incurred loss during last three consecutive financial accounting years ending the immediate last financial year in which RFP/NIT is invited, **(vi)** Should have not been in default to any banker/financial institution during last eighteen (18) months (say) ending previous day of last date of submission of proposal, **(vii)** should not be under liquidation, court receivership, or similar proceeding like referred to National Company Law Tribunal (NCLT), if so, or Insolvency and Bankruptcy Board, if so, etc. **(viii)** In addition, bidder are also asked to submit an affidavit of not being black listed by any client in last three preceding year or the bidder or **(ix)** its Proprietor/Partner (s) / Directors (s) should not have been convicted by a court of Law for an offence involving moral turpitude in relation to business dealings during the past seven years (say), etc. Failure to meet the PQ/eligibility criteria will render the bid to be summarily rejected, etc.

The technical and financial eligible engineering consultant, as evaluated as per the above prescribed PQ/eligibility criteria, are shortlisted and their techno-commercial proposal are considered for further evaluation based on the philosophies prescribed in the RFP. The evaluation method usually follow any of the following as considered appropriate depending upon the nature of assignment : (i) Cost Based System (CBS) (ii) Quality and Cost Based System (QCBS) (iii) Quality Based System (QBS) (iv) Combined Quality Cum Cost Based System (CQCCBS), etc.

CBS method: The CBS method of evaluation for selecting engineering consultant is used for routine assignments of low value, where any experienced and eligible consultant with similar past experience can deliver the service without requirement of specific expertise. The eligibility criteria are kept simple for extensive participation. Under this philosophy, the technical proposal of bidders are opened and evaluated for meeting compliance with TOR. The bidders/engineering consultant (s) having no variance/deviation with TOR are shortlisted and their financial proposal is opened. The bidder/ consultant who has submitted the lowest financial bid (L1 bidder), is selected for the assignment.

The assignments under CBS method of evaluation usually include third party Inspection services, preparation of project profile/concept notes, market survey, site survey and site selection, raw material linkage studies, conducting energy audits, preparation of Techno-economic Feasibility Report (TEFR), preparation of Environmental Impact Assessment (EIA)/ Environmental Management Plan (EMP) Report, ISO-9000 Quality Management System (QMS) Implementation, ISO-14000 Environmental Quality Management System (EQMS) Implementation, OSHA-18001/ ISO-45001 Occupational Safety and Health Administration Standard implementation, software design and development, undertaking jobs on unit/item rate contract basis like conducting environmental monitoring audit services, etc.

QCBS method: The QCBS method of evaluation is used for assignments/jobs of moderate value for technical projects under semi-routine circumstances intermittent in nature, where experienced consultant with similar past experience can deliver the service with requirement of specific expertise. The eligibility criteria are lenient for larger participation. First, bidders/consultant's technical proposal is evaluated strictly as per the information furnished in the Standard Formats of NIT/RFP to determine the technical suitability and score. The bidders/ consultants who score minimum prescribed technical qualifying marks or more are considered for opening of their price bid.

Under this philosophy, the technical evaluation criteria mentioned in RFP are divided into sub-criteria, say experience of the firm (say 40% weight), CVs of key personnel (say 40%), methodology of execution (say 20%), etc. total being 100%. A minimum qualifying marks for the technical proposal is prescribed in RFP (say 70%). All the eligible bidders/consultants who score the minimum technical qualifying marks say 65% or more, are considered for next stage and their financial proposal is opened. The bidder/consultant whose financial proposal is lowest (L1) is selected for the assignment.

The assignments under QCBS method of evaluation usually include preparation of Detailed Project Report (DPR), product mix selection studies, Clean Development Mechanism (CDM), restructuring and turnaround studies, residual life assessment studies, health studies, due diligence report, asset valuation of plant, detailed engineering services for the facilities, procurement engineering services, etc.

QBS method: The QBS method of evaluation for selecting engineering consultant is used for circumstances, where specific expertise is required or the assignment is very complex or highly specialized or technically critical projects of strategic importance, or where it is difficult to define scope of services with accuracy, or where the outcome of the assignment will have high impact and hence it is essential to engage most qualified experienced consultant with identical past experience. The selection criteria are stringent and highly pin-pointed for meaningful participation. Under this philosophy, the technical evaluation criteria mentioned in RFP are divided in to sub-criteria, say experience of the firm (say 50% weight), CVs of key personnel (say 50%), etc. total being 100%. A minimum qualifying marks for the technical proposal is prescribed in RFP (say 90%). The consultant who scores the maximum technical qualifying marks i.e. technically highest scored qualified bidder (T_1) is considered further and only its financial proposal is opened and is selected for the assignment irrespective of the fees quoted by him.

The assignments under QBS method of evaluation usually include technology supervision services, designer's supervision services at manufacturer's works and client's site during construction, erection, commissioning and conducting of performance guarantee tests, etc.

CQCCBS method: The CQCCBS method of evaluation for selecting engineering consultant is generally used for large projects of high value or projects technically complex in nature, or projects of national eminence, etc. where experienced consultant with similar past experience can deliver the service with requirement of specific expertise. The eligibility criteria are package/scope of services specific, but for wider participation.

In this philosophy, consultant's technical and financial proposal are taken into consideration in combined/synchronous manner rather than reclusively or exclusively. However, the quality of technical proposal is paramount over the financial proposal (fees/price) in evaluating the engineering consultant. First, bidders/consultant's technical proposal is evaluated strictly as per the information furnished in the Standard Formats of NIT/RFP to determine the technical score (S_t). The bidders/ consultants who scores minimum prescribed technical qualifying marks or more (S_t) are considered for opening of their price bid. Weighted average value is calculated for arriving at overall combined ranking of the technical and financial proposals. As technical proposal is paramount, a higher weightage is allotted to it (say 80%) over financial proposal (say 20%) for evaluating the proposal of bidders/ consultants.

The technical evaluation criteria indicated in RFP are divided in to sub-criteria, say experience of the firm (say 35% weight), CVs of key personnel (say 45% weight), methodology of execution (say 15% weight), financial capability (turnover from similar consultancy business, say 5% weight), etc. total being 100%. Consideration of these criteria for evaluation is to ensure the quality of service to be rendered by engineering consultant is of unwaveringly prolific standard. A minimum qualifying score for the technical proposal is prescribed in the RFP (say 75 marks) which is required for qualifying technically and for consideration of opening of financial proposal (Part-2). The financial scoring (S_f) of consultant(s) having scored 75 marks or more technically, are carried out as an inverse percentage of the lowest proposal (i.e. $S_f = L_{low}/ L$, where L is evaluated bid price of the bidder and L_{low} is the lowest of the evaluated bid prices among responsive bids.

Consultants' proposals are finally ranked according to the combined technical score (S_t) and financial score (S_f) using the formula $S = S_t * T\% + S_f * F\%$, where T is the weight given to the technical proposal (say 80%) and F is the weight given to the financial proposal (say 20%); T + F = 1. The bidder/consultant scoring the highest combined technical and financial score is selected as successful bidder and are invited for award of contract/ assignment. However, in the event of tie at the top position between two or more bidders, the bidder whose technical score is highest (more technically competence) is selected as preferred bidder. However, the preferred bidder is required to match the lowest price bid among the bidders forming the tie. If the preferred bidder does not agree to match the lowest price bid among the bidders forming the tie, the bidder having 2nd highest technical score (T_S) is considered for the award in similar manner with the lowest price bid among the bidders forming the tie.

The assignments under CQCCBS method usually include preparation of Detailed Project Report (DPR) for mega projects, restructuring and turnaround studies of large corporate house having multi location production/business units, comprehensive detailed engineering services for the total project, contracting and procurement engineering services for mega project, project management; monitoring; construction management and site supervision services, etc.

For projects funded by multilateral financial institutions like World Bank (WB), Asian Development Bank (ADB), Asian Infrastructure Investment Bank (AIIB), European Bank for Reconstruction and Development (EBRD), African Development Bank (AfDB), etc., project authorities /clients usually follow this philosophy in evaluating, selecting and appointing engineering consultants.

A typical illustrative example for evaluating and selecting engineering consultants based on CQCCBS method for engineering, procurement and construction management (EPCM) consultancy services for (say) raw material conveying system is presented in **Table 1, Table 2 and Table 3** below. The criteria and the respective marks as shown in the tables are indicative only for the purpose of illustration and may vary from project to project. The financial rating of technically qualified engineering consultant is carried out as an inverse percentage of the lowest proposal. The bidder /consultant scoring the highest combined technical and financial score is invited for award of contract/ assignment.

Table 1 – Technical Evaluation Criteria (S_t)

Sl. No.	Evaluation Criteria	Marks Allotted	Bidder$_1$	Bidder$_N$
A.	**For the Consulting Firm's Credential and Experience**	**35**		
1.	Number of Projects completed by the consultant as EPCM Consultant during last 7 years (3 marks for each completed project, subject to maximum 15 marks)	15		
2.	Specific experience of the consultant related to the assignment (i.e. Number of Projects completed by the consultant as EPCM Consultant during last 7 years for (say) conveying system for any raw material (ore) having a capacity of 2000 TPH (20 marks for two or more completed project and 15 marks for one completed project subject to maximum 20 marks)	20		
B.	**For Key Personnel (People)**	**45**		
1.	Project Manager (Must be graduate in Engineering in any discipline)	20		
	Number of Year of experience in any consulting firm (01 mark for each completed year. More than 6 months shall be rounded off to one and less than 6 months to zero subject to maximum 10 Marks)	10		
	Number of projects completed as project manager for raw material conveying system having capacity of 2000 TPH (05 marks for each completed project subject to maximum 10 Marks)	10		
2.	Key Personnel for Process (Must be graduate in Mechanical / Electrical Engineering with experience in Similar Project)	05		
	Number of Years of similar experience in any consulting firm (0.5 mark for each completed year of experience. More than 6 months shall be rounded off to one and less than 6 months to zero subject to maximum 03 Marks)	03		
	Number of completed Projects associated with (1 mark for each completed project subject to maximum 02 Marks)	02		

3.	Key Personnel for Mechanical (Must be graduate in Mechanical Engineering with experience in Similar Project)	05		
	Number of Years of similar experience in any consulting firm (0.5 mark for each completed year of experience. More than 6 months shall be rounded off to one and less than 6 months to zero subject to maximum 03 Marks)	03		
	Number of completed Projects associated with (1 mark for each completed project subject to maximum 02 Marks)	02		
4.	Key Personnel for Civil /Structural (Must be graduate in Civil Engineering with experience in Similar Project)	05		
	Number of Years of similar experience in any consulting firm (0.5 mark for each completed year of experience. More than 6 months shall be rounded off to one and less than 6 months to zero subject to maximum 03 Marks)	03		
	Number of completed Projects associated with (1 mark for each completed project subject to maximum 02 Marks)	02		
5.	Key Personnel for Electrical (Must be graduate in Electrical Engineering with experience in Similar Project)	05		
	Number of Years of similar experience in any consulting firm (0.5 mark for each completed year of experience. More than 6 months shall be rounded off to one and less than 6 months to zero subject to maximum 03 Marks)	03		
	Number of completed Projects associated with (1 mark for each completed project subject to maximum 02 Marks)	02		
6.	Key Personnel for Finance (Must be Cost Account/ Chartered Account /MBA(Finance) with experience in industry/ project)	05		
	Number of Years of similar experience in any consulting firm (0.5 mark for each completed year of experience. More than 6 months shall be rounded off to one and less than 6 months to zero subject to maximum 03 Marks)	03		
	Number of completed Projects associated with (1 mark for each completed project subject to maximum 02 Marks)	02		

C.	Adequacy of the proposed work plan and methodology in responding to the terms of reference (TOR)	15			
D.	Financial Capability : Consultancy Turnover (Consultancy turnover for last 3 years: Bidders having average annual turnover equal to or more than US\$ 30 Million (say) shall be awarder full marks. Bidders whose average annual turnover is less than US\$ 30 Million shall be entitled to proportionate score with reference to US\$ 30 Million)	05			
	Total Technical Score (S_t)	100			

Table 2 – Financial Evaluation Criteria (S_t)

Evaluation Criteria	$Bidder_1$	$Bidder_2$	$Bidder_3$	$Bidder_N$
Fees / Price Quoted (US \$ Million)	F_1	F_2	F_3	F_n
L_0 Value (US \$ Million) (Lowest quoted fees)	$L_0 = F_1$ (say)	$F_2 > F_1$	$F_3 > F_1$	$F_n > F_1$
Equivalent Financial Score (S_f)	$Lo/F_1 = 100$	Lo/F_2	Lo/F_3	Lo/F_N

Table 3 – Combined Score (Technical and Financial: S)

Evaluation Formula	$Bidder_1$	$Bidder_2$	$Bidder_3$	$Bidder_N$
$S = S_t * T\% + S_f * F\%$				
Status (Successful / Not Successful)				

Illustration:

Technical Evaluation Score (S_t)

S. N.	Evaluation Criteria	Marks Allotted	Bidder B_1	B_2	B_3	B_N
	Firm's Credential and Experience	35	29	32	33	25
	For Key Personnel (CV of Engineers)	45	37	40	36	30
	Adequacy of the proposed work plan and methodology in responding to the TOR	15	11	12	13	5
	Financial: Consultancy Turnover (Last 3 years)	5	3	5	4	2
	Total Technical Score (S_t)	100	80	89	86	62
	Qualifying Status (Minimum Score = 75)		Yes	Yes	Yes	No

Financial Evaluation Score (S_f)

Evaluation Criteria	Bidder$_1$	Bidder$_2$	Bidder$_3$	Bidder$_N$
Fees / Price Quoted (Million US $) say	10 (L_0)	11	12	*Not Opened*
Equivalent Financial Score (S_f)	100	91	83	-

Combined Score (Technical and Financial: S)

Evaluation Criteria	Bidder$_1$	Bidder$_2$	Bidder$_3$
$S = S_t * T\% + S_f * F\%$	$S = 80 * 0.8 + 100 * 0.2 = 84.0$	89.4	85.4
Status	-	*Successful*	-

Conclusion: Bidder$_2$ having technical edge bags the job though fees quoted is 10% higher than Bidder$_1$.

As an alternative, some Clients do calculate evaluated bid score for each bid, which meets the minimum technical qualifying marks (say 75 marks), using the following formula in order to have a comprehensive assessment of the bid price and the quality of each bid. Consultants' proposals are finally ranked accordingly using the combined technical score (S_t) and financial score (S_f). However, the net outcome, i.e., ranking of the bidders remains unaltered w,r,t. the formula used above.

$$S = S_t * 100 * T\% + S_f * 100 * F\%$$

$$\text{Or, } S = (T/T_{high}) * 100 * T\% + (L_{low}/L) * 100 * F\%$$

Where,

T = Total technical marks obtained by the bidder against Quality Criteria

T_{high} = Total technical marks achieved by the best bid among all responsive bids against Quality Criteria

L_{low} = The lowest of the evaluated bid prices among responsive bids

L = Evaluated bid price of the bidder

T = Weight given to the technical proposal (say 80%)

F = Weight given to the financial proposal (say 20%) Or (1- T)

Chapter 8 – Defining Engineering Consultancy Marketing

For service industry, it was observed that the traditional marketing mix as coined by Professor Neil Borden (twelve elements) and summed up by McCarthy into 4Ps namely product, price, place and promotion were inadequate. Booms and Bitner [10] have suggested an expanded marketing mix model. Magrath [11] endorsed such an approach and categorically pointed out that when marketing services, 4Ps are not enough. He said that another three Ps, namely personnel, physical facilities and process management must be included in the marketing mix. A number of marketing research studies supplement the relevance of each of the 7Ps. Like, Shostack [12] gave due emphasis on physical evidence. Cowell [13] identified the issues in process management. Judd [14] came out with the importance of people. Following the trend, Agrawal [15] expanded the mix of marketing elements to eight variables by including a very important P i.e. pace. Agrawal suggested that the pace of the marketing response is controllable by marketers and therefore is to be included in the marketing response as are the rest of the marketing elements.

Bearing the above in mind, marketing of engineering consultancy can be defined as the process of identifying client and their changing techno-economic project requirements, converting them into required products/ services, making them available at convenient places, pricing them reasonably (charging fee), communicating (promoting) with them and initiating action (sales), rendering services with pace, execution through right set of people and proper IT supported process backed by required physical infrastructure, so that the process of repeat order continues through client delight and competitive differentiation. In this definition, marketing research assumes the task of identifying clients and their changing techno-economic project requirements, while the marketing mixes **(8Ps)** are used to deliver solution. Thus, engineering consultancy marketing commences much before the services are rendered and continues after the services are over. To succeed in this marketing warfare, it is essential for engineering consultancy firms to know the markets/ clients, their changing requirements and the competitive forces as well. This calls for market acquaintance, understanding the clients and customization of the services. Thus, engineering consultancy marketing is a three tier system (3S) comprising of searching/ selecting the markets/ clients, serving the selected/identified markets/ clients and sustaining the market/client through client relationship marketing (CRM) and key account management (KAM).

Chapter 9 – Need For Marketing System For Engineering Consultancy

Drucker [16] has mentioned that there are basically the two essentials for any successful business: innovation and marketing. He further said that a company's first task is to create customers. Levitt [17] in his classic *Marketing Myopia* emphasized the need for farsighted marketing for firm's growth. He remarked that most businesses suffer from myopia when they fail to ask themselves a question, *what business are we in?* Levitt suggested that firm must learn to think of itself not as producing goods or services, but as producing followers i.e. the customers/clients. To produce these customers/clients, the entire corporation must be viewed as a customer–centric and customer-satisfying organism. The firm must aim at acquiring customers by taking pains and walking that extra mile and offering the things that will make them do business with it. Levitt's arguments were expanded by Peters and Waterman [18] in their book *In Search of Excellence*. They emphasized that corporate culture based on innovation and closeness to the customer is more important. They concluded that excellent companies really are close to their customers. That's it. Other companies talk about it; excellent companies do it. In the above back drop, some of the specific reasons entailed for marketing management system for the engineering consultancy industry are expatiated below :

- For identifying and selecting clients who can be served better and profitably.
- For understanding client's techno-economic requirements before hand as every client prefers to discuss the project and scope of

work with engineering consultant who is going to execute the project.

- Just doing good basic engineering and detailed engineering work won't assure further business, either with the same client or others especially in present situation where today's differentiated offerings becomes tomorrow's quintessential requirement and there are many firms willing to render the same. A close situation monitoring is required as it helps in getting insight about the client, their changing techno-economic requirements, the competitors' offerings, etc.
- For understanding the market dynamics, scanning of opportunities and threats, selecting the target market, designing the customized marketing mixes and realigning the firm's scarce and costly resources to the target market/client best suited to increase marketing productivity.
- For organization promotion, as no engineering consultancy firm automatically becomes known because it is competent in its profession.
- For service promotion, as a superior service does not sell by itself. Its differential merits and tangibilities must be communicated well to the industry/market/ client.
- Educating client as they have imperfect/ perplexing information due to 'n' number of consultants willing to render similar services. An engineering consultant acts as an advisor/ counselor and assists the client in decision making during project conceptualization stage by defining the project scope, capital investment required, agencies/ contractors to be involved, risks associated, remedial measures, explaining the benefits client will have from the use of services from the consultant etc.
- For marketing intelligence as every firm has to have information on what is going on in the investment front, engineering and technology front, market, competitive forces etc.
- To have competitive edge by minimizing/ nullifying/ pre-empting threats at the earliest opportunity.
- For effective implementation of relationship marketing programs, as contact between the firm and the client is important for mutual understanding, trust, delight and patronage.

Chapter 10 – Members of Engineering Consultancy Marketing Team

The unique features of engineering consultancy services make marketing philosophy different as compared to industrial product marketing or capital goods marketing. In case of industrial products or capital goods, the marketing function is mainly handled by marketing department alone. Here, as far as customers are concerned, marketing department can plan and implement most of the marketing activities and can control almost the total marketing functions.

But in case of engineering consultancy, the situation is entirely different. The engineers who work for the project are perceived to be an integral part of the services by the client and their project team. Client expects that consultant's engineers should be willing to walk that extra mile and act as a bridge between various department of both the firms, rather than concentrating on one to one stand alone relationship with a particular department issuing the assignment. An engineer of engineering consultant markets itself every time a client's personnel interacts with him and every such interaction creates *moment of truth*, about the consultant and provides /withholds marketing opportunities. For example, a techno-economics engineer is engaged in marketing while explaining the economics of the project as calculated by him in the techno economic feasibility report (TEFR). If he is not persuasive, then, as far as the client is concerned, the engineering consultancy firm is not persuasive. The question is not whether the techno economic engineer will market the consultant's services efficaciously, or he should engage in marketing, marketing is immanent in his job. Effective integration of marketing concept throughout the firm increases the likelihood that the job of every member of the team will be performed in a manner consistent with the marketing concept. Thus in addition to personnel belonging to marketing department, each and every engineer of the engineering consultant, irrespective of the discipline or department they belong to, is a man of marketing in a subtle way. Further, every engineer of consultant is member of some professional body and has contacts with many fellow members. From them he may learn of potential client, or have information on what is going on in the market and investment front, etc.

In engineering consultancy, every engineer is the brand ambassador of the firm and by default is involved in prospecting task (the first and critical milestone of marketing). In engineering consultancy, brand is built by all the engineers whereas branding is by few personnel of corporate communication department. Brand building in engineering consultancy is a voyage of building a corporate soul and diffusing it inside and outside the firm to all stake holders namely clients in general and key clients in particular, employees, technology suppliers, Vendors, Government, Regulatory bodies, the surrounding around, etc. In engineering consultancy, the power of brand building shouldn't be taken lightly nor should it be considered as another fad which would pass away like many earlier innovative management concepts. The truth is: brands appreciate more than land, gain better returns and grow faster and at times exponentially giving 'multi-bagger' returns.

The information thus gathered by the front end engineer should be passed on to marketing department for carrying out further marketing efforts. Gummesson [19] defined such employees as *part time marketers*. He opined that they are the people who are not on the roll of a marketing (or sales) department, whose main task is something other than marketing and who normally are not trained in marketing skills but who, when performing their tasks, directly or indirectly influence quality perception, client delight and client's future purchase decisions. He pointed out that if the part time marketers are not prepared to play/take their role in marketing, the value process will be maimed and the marketing endeavor will be less effective and will fall short to the targeted dividend regardless of how good the persuasion process is, and perhaps may crumble and get obliterated. The onus of creating the *moment of truths*, multitude of contacts and business leads cannot be solely vested on an individual like General Manager (Marketing) or Director (Marketing) of the firm.

Marketing should be in the blood and breath of each and every employee of the engineering consultancy firm and not just in the marketing department personnel. Marketing must become the business philosophy and concern of the entire firm. Thus, in engineering consultancy, marketing is too important a task to be left alone to the marketing department. The role of non marketing department personnel is vital. For marketing success, engineering consultants must understand these aspects well and the manner in which they impinge on marketing strategy formulation. To this end, engineering consultant should market firm's marketing philosophy internally to the employees first (internal marketing) and then go to the external clients. To measure up to the business realities, to cope up with the paradigm shifts and to realize its potential in external marketing, it is imperative for the engineering consultancy firm to stress on internal marketing and position 'concern for quality' and 'concern for client' among all the engineers (i.e., the front line Client contact engineers as well as engineers who are rendering services behind the scene). Effective integration of marketing throughout the firm will augment the likely hood of performance enhancement of every engineer in a manner consistent with the firm's marketing objectives. The information gathered by the front end engineers should be passed on to the marketing department for carrying out further marketing efforts. However, the role of such part time marketers (engineers) should be restricted to prospecting, sharing order bagged news, making technical presentation/demonstration with solution selling approach, clarifying the issues, etc. thereby creating a business lead and carving a niche. Rest of the marketing activities (sales activities) involving offer submission, negotiation, order finalization, etc. should be taken care solely by the marketing department as it is their line of business (LOB). Company information like major order bagged, project commissioned, time schedule, etc. available in public domain should be consider as shareable information and not confidential information. The part time marketers (engineers) should use such information/figures (duly rounded off) for creating marketing lead. But, they should never indulge in sales activities as mentioned above. Issues like modus operandi (operating from engineering office or client's site), expected fees, man hour charges or man day charges or man week charges or man month charges, etc. which are sacrosanct should not be discussed by them with the prospect/client at any cost.

Chapter 11 – Internal Marketing in Engineering Consultancy Services

The relevance of internal marketing in service sector is undebatable and well established in developed countries. It is no longer a buzzword, but necessity for every service business including engineering consultancy. As economies are becoming service oriented, its practice has attracted considerable interest among the service firms around the world, especially in the United States, the United Kingdom and the countries in Nordic region. There is an increasing concern for internal marketing in service industry namely health care, hospitality, tourism, financial services etc. Many firms operating in these sectors have made sufficient progress. However, in case of engineering consultancy, the progress achieved is not so as expected. There are various reasons. The few such reasons which act as barriers are sales oriented mindset i.e. firm's key concern is for immediate sales rather than relationship building, firm wide lack in appreciating the criticality of the role engineers play in service transaction, business values of some clients, indifferent attitude of some clients towards consultant, feeling among the employees that reward would favor for procuring immediate orders from the client rather than long term gains, attitudinal barriers say *'keep going'* attitude of employees, etc. In fact, in engineering consultancy, practice of internal marketing is more essential due to aspects like limited client base and strategic dependence on a few key client, co-existence of service provider and service receiver, people factor, quality dimensions, etc. In view of these and to realize its potential in marketing, engineering consultancy firms need to stress in making internal marketing happen. The objective is to develop competent, motivated, satisfied and customer conscious work force who is willing to walk that extra mile with the customer/client.

11.1 Client Relationship

As depicted in **Figure 1,** the co-existence of consultant's design and site engineers and client's project engineers heightens the need for strong client concern and relationship. In engineering consultancy, ongoing relationship with client is a major business asset. A healthy relationship helps the consultant in two ways. Firstly, it helps the consultant in securing fresh assignments/ work orders from the client on nomination basis. In fact, majority of engineering consultancy jobs are awarded based on the consultant's past performance and the client's satisfaction with them. Secondly, it helps the consultant in rendering services in a quality manner. As, for an engineering consultant, the client is also a source of human resource, especially in execution of mega projects where concurrent design and engineering, plant and equipment supply, civil construction, site erection, testing and commissioning take place. The quality of service depends on the client's project team's attitude towards the consultant. Client's project team having positive attitude cooperates well with the consultant's team, takes prompt decisions, invests time and resources in describing its requirement, furnishes techno commercial and site information in time and accurately, makes the front available when required, etc. As the client's involvement is a requisite for successful implementation of project, without the client's involvement, the quality of engineering consultancy service may be far from usurp. Thus, in engineering consultancy, the client – consultant relationship does not end when a sale is made/ order is procured. In fact, the relationship intensifies after the work order is received and continues till project completion and subsequent receipt of fresh assignments the next time. The challenge before an engineering consultant, is therefore, to make the organization client centric and to enjoy client's confidence and patronage. This justifies that the practice of client relationship management is highly essential for engineering consultancy [20].

11.2 The People Factor

The importance of people in the marketing of services is well proven. Authors like, Sasser [21], Judd [22], Stershic [23], Bartlett and Ghoshal [24], Parasuraman [25], and many others have opined that in the service industry employees are an integral part of the product itself. The role of contact personnel (front line employees) is important for service firms in delighting customers. Services personnel perform the dual functions of interacting with the firm's external environment and the firm internal and this role is referred to as "Boundary Spanning" role (Srinivasan [26]).

For clients, the service is marketed, produced and delivered by the people. They spelt out that a service company's identity is through its people. Since a service is a performance performed by the people, it is difficult to separate the performance of the service from the people performing the service. If the people don't meet customer's/client's expectations, then neither does the service. Thus, people play a vital role in marketing of services and hence it warrants its inclusion as a distinctive element of the service marketing mix. For service firms who are in intellect business, it is human capital, not financial capital, which should be the starting point for formulating business strategy.

The same is true for the engineering consultancy sector also. In engineering consultancy, people/employees/ engineers are the face of the consultant/firm, their knowledge, their competence, their skill, their expertise, their attitude, personal traits, courtesy, etc. are consultant/ firm's USP (Unique Selling Propositions). An engineering consultant's identity in the industry is established through its engineers. They are perceived to be an integral part of the service offerings as being directly involved in service production and service delivery. They give cutting edge in the market place resulting in repeat orders from clients.

Engineers of various disciplines like technologists, process engineers, metallurgical engineers, civil engineers, structural engineers, electrical engineers, instrumentation engineers, automation/ IT engineers, utility engineers, estimation engineers, techno-economics engineers, contracts engineers, construction planning engineers, inspection engineers, construction and erection engineers and other site engineers constitute the central force/nucleus in the marketing mix. It is the engineers who ultimately create the service differentiation. They manage the *moments of truth* with the client's project team, contractors, vendors, statutory authorities, etc.

In fact, the consultant's engineers are the product and hence the most valuable asset. Clients view the engineering consultant/ firm through the disposition of these engineers. In their eyes, these engineers who are either at the front line or supporting the front line from behind the scene, are the product, service and the firm. Only when these engineers of the consultant deliver the services in a quality way, with utmost concern for the client and successful completion of the project without time and cost overrun, does the likelihood of customer delight, patronage and placing additional responsibilities increase. Thus, the success of engineering consultancy service is tied closely to the management of quality people (engineers) as it is the people element (engineers), which ultimately makes the difference.

11.3 Making Internal Marketing Happen in Engineering Consultancy

There is no commonly accepted definition of internal marketing. Several gurus/ authors/ have expressed their opinion on internal marketing. As per Sasser and Arbeit [21], internal marketing starts from a notion that a service firm in order to be successful must first sell its business/marketing objectives to its employees in order to be able to sell its products / services to the customer. The objective of internal marketing is, first, to employ and keep the best people and second, to make them do the best possible job by applying the philosophy and practices of marketing internally. The concept holds that an organization's internal work force can be influenced most effectively and hence motivated to customer/client consciousness and marketing mindedness by adopting and applying marketing like activities internally. The employees should be motivated and pursued to react in the desired way, just as customers on the external market are motivated and pursued to respond favourably to the firm's offerings. Gronross [6] opined that the marketing activities in service organizations should be broader than the traditional marketing realm. He further said that marketing activities in service organizations can be clubbed under three major heads, namely: internal marketing, external marketing and interactive marketing. **Figure 2** depicts Gronross's three stage model of external marketing, internal marketing and interactive marketing. It is Parasuraman [25] who modified the model by introducing a new dimension 'technology', as an enabling and facilitating tool which provides a supportive environment for service production and delivery.

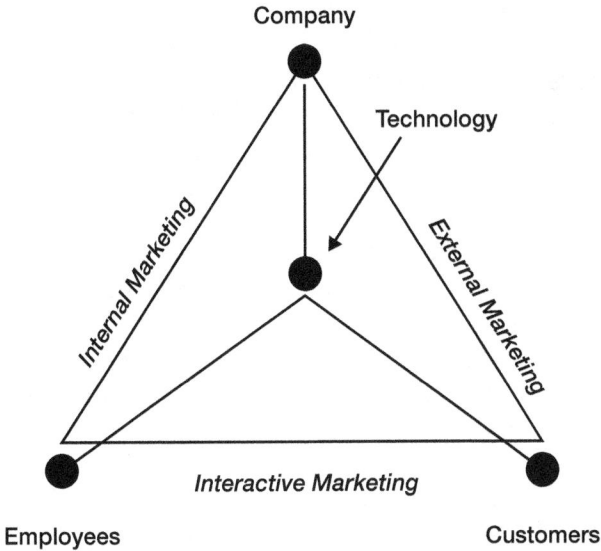

Figure 2 – Gronross's Services Marketing Model.

Internal marketing is customer/client oriented approach whereby a firm seeks to create customer/client consciousness and concern among its employees. Its central tenet is the creation of client centric organization. Xavier [27] suggested that internal marketing aims in germinating market oriented and customer focused culture in an organization. It treats the customer contact employees as internal customers and keeps their morale and motivation high. It creates an environment where customer contact employees feel motivated and empowered to deliver high quality services to customers. Zeithaml and Bitner [28] have mentioned that through its external marketing efforts, a company 'makes promises' to its customers regarding what they can expect and how it will be delivered. Promises made must be kept. 'Keeping promises', or interactive marketing, is the second type of marketing activity and is most critical from the customer's point of view. Service promises made by the external marketing employees are kept or broken by the internal employees of the service firm. Interactive marketing occurs in the moment of truth when the customer interacts with the organization and the service is produced and consumed. In reality, promises are easy and trivial to make but onerous to deliver.

For delivering promises, a third form of marketing called internal marketing takes place through the 'enabling of promises'.

Parasuraman [29] further opined that, in the case of services, the key to achieve a competitive advantage is by serving customers through a fine blend of internal marketing (marketing to employees) and interactive marketing (employees marketing to customers) rather than merely marketing to them (external marketing). Expressing similar opinion, Barnes [30] mentioned that if the staff would not like to buy it then why should the customer? Green et al [31] opined that Internal marketing is the key to external marketing success. Only a satisfied employee can keep the customer satisfied.

Kimura [32] conducted an empirical study into the current state and structure of internal marketing in Japanese companies. The research study empirically examined the relationship between the degree of penetration of internal marketing and corporate performance in Japanese companies. The research finding indicated that internal collaboration between functions and operational strength are the two formidable main factors for the realization of firms' market development. Furthermore, internal collaboration was particularly influential in the firms which successfully conducted external marketing.

For engineering consultancy sector, practicing internal marketing is not a choice but an essential requirement as it paves the way for successful external marketing. In order to deliver on the commitments made and achieving competitive advantages, engineering consultants need to employ the best engineers and retain them. Since engineering consultancy is a performance and the performers are the engineers, delivering quality service consistently and managing moments of truth require that the engineers are carefully cast (recruitment and selection), well choreographed (training), fully cognizant of how their roles interrelate (teamwork) and provided with the necessary support (incentives and rewards). The business challenge for engineering consultancy firm is not to build client loyalty, but, in fact the reverse i.e. to build employee's loyalty to the client. Clearly, the onus of instilling the client's concern among internal employees/ engineers lies with the firm.

From engineering consultancy perspective, the interactive marketing phase is vital. During this phase, the interaction between the consultant's engineers and client's project team is maximum. Due to longer duration of interaction between the consultant's engineers and the client's project engineer (because of long gestation period of project execution), the need to sensitize engineers to the importance of client delight becomes more important. The engineers of various departments and disciplines, as mentioned earlier, need to know that they are involved in marketing activities as part time marketers. They should be made clear that if they do not meet the client's expectations, neither does the firm. That, if they are not serving the client directly, they are serving someone who are serving the client. They must be impressed upon that teamwork is the heart of success. That, engineering consultancy is like track-and-relay event where no runner/ engineer runs/ works alone, and each one actually provides some service to the other in the service delivery chain. All are part of the same team, yet the second runner's ability to run a good race and make a successful hand off to the next member of the relay team depends on how smoothly the first runner passes the baton to him. These hand-offs affect not only the next leg of the race, but the success of the entire team. To make sure that the hands-off are smooth and effective, they must work as an integrated team. Instead of competing, they should compliment and co-operate each other.

Figure 3 depicts engineering consultancy process and interrelationship of various departments for a typical work situation [33]. Upon receipt of an engineering consultancy work order/assignment from the client (say setting up of a utility power generation plant), top management nominates a project coordinator. The project coordinator prepares project specific quality plan (PSQP) and distributes the document to all concerned technical sections likely to be involved in execution. This project specific quality plan lists out the project profile, scope of work of the technical sections, execution schedule, specific practices to be followed etc. The project coordinator issues primary engineering assignment to technology group (Power Generation Group). The technology group issues secondary assignment to services groups (Coal Handling and Preparation, Utilities/Fuels and Gases, Hydro Engineering, etc.), engineering groups (Civil, Steel Structural, Power Distribution and Shop Electrics, Material Handling, Environmental, etc.) and other groups (General Layout and Transportation, Fire Protection and Safety, Construction Planning and Monitoring, Procurement Engineering etc.). These groups exchange assignments among themselves under intimation to the Project Coordinator as shown in **Figure 3.** Deliverables like assignment drawings, package specific technical specifications, calculations, schematics, assignments for further engineering work, etc. are generated by the respective sections and is passed on to the

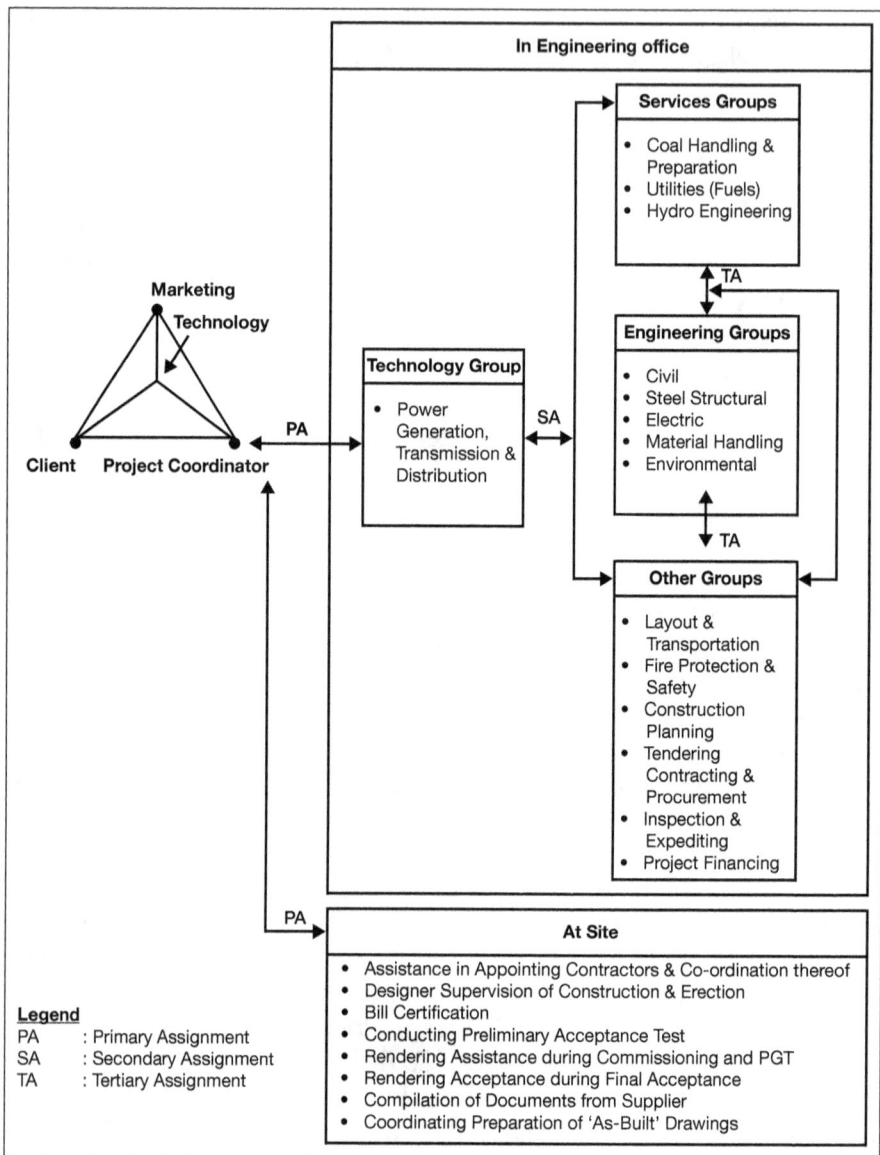

Figure 3 – Engineering Consultancy Process (Typical Work Situation).

project coordinator/ concerned section depending upon the nature of the requirement. Once the engineering activities are over, civil construction, plant and equipment procurement orders are placed to various agencies/ contractors, and then the site activities start.

Thus, every engineer and every engineering, technology and services department within the organization have roles as internal customers as well as internal suppliers. To maintain high quality service, every individual engineer and every department needs to work together in a well knit manner in a way that is aligned towards a common goal i.e. successful commissioning of the project, meeting client's expectation and delighting the client.

Thus, in order to be successful in its external marketing endeavor, internal marketing to the internal employee/ engineers is essential. An engineering consultancy organization should first treat its engineers as internal customers and sell firm's business philosophy internally to them. For engineering consultancy, if the external marketing is for understanding client's requirements and expectations, then internal marketing is to penetrate the external marketing substance for acting as an input for interactive marketing.

11.4 Putting Internal Marketing in Action

Despite its name, internal marketing is more a management strategy rather than a marketing function. Gronross [34] opined that internal marketing should be viewed as a managerial philosophy, which has strategic and tactical implications throughout the firm and its various business functions. For successful implementation, two basic components namely 'People Management' and 'Communication Management' needs to be stressed upon. The ten dimensions of service quality are influenced directly by these components. No unique initiative will yield dividend unless these have the right foundation. And the right foundation is total quality people (TQP) and sound communication system.

TQP (i.e. people with integrity, good values, character, and a positive attitude) management deals with getting the right set of people, retaining them and satisfying them to buy into corporate goals. Whereas communication management deals with managing the information that employees need to perform effectively and discharging their responsibilities. In this context, internal marketing can be viewed as an umbrella concept encompassing a range of internal activities to be undertaken in the management of people and communications.

The most important areas within people management are recruitment and selection, training and development, compensation, empowerment, motivation, recognition, career path, transparency of working, job satisfaction and employee retention. The organizations must recruit people/ engineers who meet the right profile in terms of competence, skills and attitude and provide right working atmosphere to achieve high quality service. Engineering consultancy firms must realize that employee satisfaction and client satisfaction are inextricably linked. Employee satisfaction is essential as it act as a stimulus for committed performance. Only a deeply satisfied employee can keep the client deeply satisfied. It is the $E=mC^2$ of customer/client loyalty. The engineers who are actually going to render the services and interact with the client during service delivery phase, need to be quality people. They should be treated exactly as the firm wants them to treat clients. This implies that considerable attention should be focused on attracting and recruiting the right set of engineers both at the initial level as well as at the lateral entry level, developing them and motivating them to take up the challenges.

Further, sound and healthy interpersonal communication among the employees within the firm is another important area. Internal communication is the foundation of all internal marketing programs. It is a vital tool for binding the organization, enhancing employee motivation, promoting transparency and reducing attrition. To be effective, it needs to travel in all directions: top down, from senior management to all the employees; bottom-up, from all levels of employees back up to the senior management; and laterally, across all levels of the firm. If communication is proper, or complete, it will reach to all those who are involved leading to high morale, motivation and satisfaction. Otherwise, due to communication gap, rumors may start circulating, worse yet, the employees may feel excluded and not a true part of the system. Therefore, an environment should be created and maintained in which employees feel informed enough to feel that they are the business and they own the business.

In engineering consultancy, proper communication is much more important due to inter-dependence of various engineering departments as discussed earlier and depicted in **Figure 3.** The ability to work together is the key attribute for successful discharge of respective responsibilities. All functional areas, be it marketing or any other, have to be equally concerned. The corridors of marketing, design and engineering, project departments, finance, contracts engineering, must be kept wide open for an effusive flow of ideas between client and these departments.

Chapter 12 – Measure to be Adopted for Increasing Marketing Effectiveness

As competition and business complexities increases, more marketing sophistication is needed. The major marketing challenge before engineering consultancy firms is to create competitive edge. To create competitive edge, engineering consultancy organizations must stress on the following: Client selectivity, Client relationship management (CRM), Appointment of Key Account Manager (KAM), Differentiated offer, Differentiated service delivery, Solution Selling/ System Selling, Industry and Competitor analysis, etc.

12.1 Client Selectivity

Market segmentation, targeting and positioning are known as the STP of Marketing. Kotler and Keller [35] opined that firms cannot connect with all consumers in large, broad or diverse markets. They need to identify the market segments they intend to serve optimally keeping operating parameters and constraints in mind. This decision requires keen understanding of the market and the customer/client buying behavior and careful strategic thinking about what makes each segment unique and different. Market segmentation divides the market into well defined slices which the firm can serve effectively. Identifying and uniquely satisfying the right market segments is the key to marketing success. Firm's task is to identify appropriate number of market segments based on characteristics like geography, demography, buying behavior, buying volume, future investment plans, etc. and asking whether these segments exhibit different needs and product/service sponsors and decide which ones to target. With above backdrop in mind, engineering consultancy market can be segmented geographically, demographically and behaviorally.

Having clustered the market in segments, the next task is to decide how many and which customers/client to target. Not all segments/customers/clients are meaningful and profitable. For targeting, the firm must look at two factors; the segment's overall attractiveness and the firm's objectives and resources. Once the market has been segmented and the target market/customer/ client identified, the next job is to position the product/service in the market/customer. No company can win if its product/services resemble every other product/service. Hence, the need is to create distinct product/service differentiation through positioning. Positioning is the act of communicating the firm's offering to occupy a distinctive place in the minds of the target market/client. The aim is to position the product/ service in the target client's mind to maximize the potential benefits to the firm. Positioning requires that the firms define and communicate similarities and differences between their product/ services and product/services of its competitors. Engineering consultants may use differentiated offer, differentiated service delivery, solution selling/system selling concept, etc. as a tailored tool to position its services in each target client's mind. The issues are covered below in this section.

In the above back drop, Kotler [36] has categorically mentioned that customer selectivity is an important facet of marketing. He opined that all customers are not worth keeping. Customers who can be turned sooner or later into a profitable accounts should be chosen and nurtured to do business with on a long term basis. After all, marketing is defined as the science and art of finding, keeping and growing with profitable customers. Not all customers are equally profitable as every one is unique in itself having differing expectations, purchasing needs, bargaining power, etc. It is not feasible for any firm to serve every customer and satisfy every customer function.

Expressing similar views Seybold [37] opines that any business initiative has to commence with targeting the right customers. It makes it easy for companies to do business with such customers and vice versa. She insisted that it is the pre-requisite for success of customer loyalty building program. In this regard, it must be clearly understood that the objective of customer selectivity exercise is not to prune the customer base but to identify the right customer mix whose propensity to be a long term business partner is high, who can be turned sooner or later into a profitable account and create value for the firm

The above view holds good more so for engineering consultancy marketing due to nature of engineering consultancy business i.e., limited client base and strategic dependence on a few key clients. Engineering consultants should target the right corporate houses/ client, the one who can be profitably served, the one who are receptive to a healthy relationship and should cultivate them, build relationship with them over a period of time to do business in the years to come. Thus, client selectivity becomes an important strategic variable for marketing of engineering consultancy services.

Porter [38] suggested selection of customer along the four dimensions: Purchasing needs versus company capabilities, Growth potential, Intrinsic bargaining power (the leverage the buyers can potentially exert over seller, given their clouts and alternative sources of supply available and propensity to exercise this power in demanding low price) and Cost of serving. Whitney [39] mentioned three criteria to classify customer: Strategic nature, Volume of business and Profitability.

With the above backdrop in mind, and taking a lead from the views expressed by the marketing academicians and practitioners, ten criteria, as shown in **Table 4** are being proposed for selecting clients in the context of marketing of engineering consultancy services. The ten criteria can be used to select clients meriting as business/ relationship partner [40]. However, one can remove/ add/ modify criteria depending upon the marketing situation.

Table 4 – Client Selection Criteria

Sl. No.	Selection Criteria
A.	Client equity.
B.	Strategic nature of association.
C.	Technical and other project related expectations and firm's performance gap.
D.	Purchase procedure followed: say simple and transparent, likelihood of awarding the contract on nomination/ repeat order basis, tendering followed by technical and financial rating of consultant, etc.

Sl. No.	Selection Criteria
E.	Bargaining power: say more demanding and discerning.
F.	Corporate culture and style of functioning like bureaucratic approach and hence very difficult/ expensive to serve
G.	Clarity of requirement (technical specifications, lot, and delivery schedule) with no midterm changes, etc.
H.	Values relationship like respect to consultant, keeps payment commitment, integrity, reciprocates and appreciates, realises vendor's genuine difficulties, willing to share views and time to tell how further improvement can be made, propensity to be loyal or switch, advocate/ supporter.
I.	Financial competence and past payment records.
J.	Long term future growth prospect.

Table 5 illustrates the method of calculating client equity, one of the most important criteria for client selectivity, based on discounted cash flow method [41]. Client equity can be increased by increasing duration of business years, increasing the range of services rendered, maximizing order procured value, achieving differentiation and cost leadership simultaneously, maximizing profit margin per unit, optimizing client acquisition cost as well as client retention cost. Cost optimization can be achieved by effective management of sales force, sales calls, use of IT, video-conferencing, instead of personal visit, use of social media, reducing marketing and promotion cost, appropriate scheme for extras, properly thought off discounts for major work orders, etc. The biggest advantage of client equity formula is that rather than saying clients are important, a firm can tell its employees the worth of each client in terms of US $ Z Million. This may help in changing the attitude and behavior of the employees.

Probable clients can be rated on these ten aspects on a Scale of 1 - 10 individually and summed up to arrive at the total marks scored as shown in **Table 6**. Client having maximum favorable magnitude on a criteria can be allotted 10 marks (say) and judgement can be applied accordingly for others. Depending upon the total scores, clients/ corporate houses can be grouped into Class – A Client, Class – B Client, Class – C Client and so on. Client / corporate houses scoring high marks naturally matter most and need the best attention. For others, the degree may vary. Such grouping will help in determining the nature of marketing endeavors and tailor made marketing strategies that need to be emphasized upon.

Table 5 – Client Equity

Client equity: The asset value of a Client relationship is called Client equity. Like any asset, the value of Client equity depends on the net earnings over the defined time period. Calculating Client equity encompasses identifying the revenue to be earned from a particular Client or group, the out flow necessary to establish and maintain the relationship and the period over which the relationship will continue. (41)
i.e., Client equity equals likely inflows from Client i.e. Client's revenue potential (with regards to core competence of the engineering consultancy firm and its competitive edge), less outflows in terms of Client acquisition cost, Client retention cost, etc.
 	
Client equity can be expressed mathematically as ,
 	
Client equity = Client revenue potential (R) – Client acquisition cost (A) – Client retention cost (Rt).
 	
$$CE = \sum_{t=1}^{N} Q_t P_t K^t - A_1 - \sum_{t=1}^{N} R_t K^t$$
 	
where,
    CE = Client equity
    R   = Client revenue potential = $Q_t\,P_t\,K^t$
    Q = Value of Order placed
    P = Expected Profit margin or mark-up
    t = Time period over which relationship will continue = 1, 2, 3, …. N years
    A = Client acquisition cost
    R = Client retention cost
    K = Discount rate = $1 / (1 + r)$
 	
Note : In case, if it has taken more than a year time to acquire a Client, which is common in Engineering Consultancy, the Client acquisition cost need to be compounded.

Preventing Death of the Concept

The idea of implementing Client selectivity in engineering consultancy marketing seems logical and obvious. Explaining the idea does not take a long time. Its principle sounds great. But, can it really work in today's cut throat competitive scenario or in depressed investment climate, where the key concern is for immediate order booking/ sales rather than having gains in the long run. Is the concept of Client selectivity practical or is it a fallacy or another fad which would pass away like many earlier management concepts?

Table 6 – Client Selection Matrix

Sl. No.	Name of the Client/ Corporate House	Criteria (Scale 1 - 10)										Total Score (100)	Client Class
		A	B	C	D	E	F	G	H	I	J		
C_1													
C_2													
C_3													
C_4													
C_5													
C_6													
C_n													

Yes, it is possible and it is essential to go for Client selectivity. In good times, Client selectivity is important, in bad times it is vital. As during the phase of recession and stiff competition, market variables and Client parameters may change abruptly. Therefore, it is essential for the firm to select the right set of Clients, do business with them and delight them.

However, there are a plethora of factors at work that destroys this nice logical philosophy. By analyzing these factors and nullifying them, the engineering consultants may minimize the risk of falling short of goals. These are, sales oriented mind set of the marketing directorate, inadequate support from the top management, poor information base, non professional attitude of the territory managers, being too diplomatic, be complacent, making too many or too few Client visits, etc. The emphasis should not be procuring more jobs quantitatively but should be procuring more jobs qualitatively.

The concept of Client selectivity is simple, but implementing the same calls for a dedicated approach that differs from those associated with traditional sales management. Its ultimate success comes from the willingness of many, from the board room to down the line, to be prepared to accept the philosophy and to mould the system so that it delivers results. The challenge for engineering consultants is to sell the concept within the organization first and find its acceptance. For that, internal marketing programs need to be conducted organization wide on a regular basis.

12.2 Client Relationship Management (CRM)

One may well question the specific relevance of the concept of CRM in engineering consultancy industry. After all, for decades practitioners have been taught that the "customer is king and is not someone to argue or match wits with", that "customer is not always right but it is often worth letting him be right". Then why has CRM gained sudden momentum and importance now?

The reasons to have CRM in engineering consultancy is increasing client defection due to post purchase dissonance complimented by competitor's offerings. Though client had always been portrayed as a 'king', but majority of clients happen to have caustic experience while dealing with the firm. Behavior and attitude of the personnel are fine till the order is placed. Afterwards, they become indifferent and usually provide 'lip' service only. The client can see through. If the attitude is not sincere, the smile proves to be irritating. How long can a client bear an irritating smile and lip service? Thus the phenomenon of post purchase dissonance creeps in and often taints the credibility of the firm. Majority of the clients defect because of the attitude of the client's contact/ front line employees. In marketing war fare, it will be catastrophic deterrent on the part of the engineering consultants to think and do so. Thus, to measure up to the realities, it is imperative for the engineering consultancy firm to have a changed mind set.

In engineering consultancy, there are many valid reasons behind the relevance of CRM. The practice for on-going relationship and intimacy with client is essential. Firstly, engineering consultancy, by virtue of its limited client base and strategic dependence on a few, can present a fertile breeding ground for CRM to flourish. Secondly, for engineering consultancy firms, acquiring and serving new clients is more expensive, cumbersome and risky, especially when there is no certainty about the present client, let alone getting client five years from now, next year, next month, or even tomorrow. As such it is more beneficial and cost effective to retain existing clients than to create new ones. Satisfied clients tend to be repeat purchasers, are more loyal, less likely to stray to competitors, and less price sensitive. The higher profits from existing clients come from large share of purchases, referrals and price premiums. The longer a client stays with the firm, greater would be the chances of obtaining large share of his business/ work orders. For engineering consultancy services, existing clients represent, by far, the best opportunities for growth. As such engineering consultants cannot afford to lose them, hence retention is the best policy. Thirdly, a close association helps in designing the customized marketing mixes because, due to their past association and understanding; the consultant is in a position to judge clients' requirements, capacity, and constraints well. Also, the client, because of his confidence on the consultant co-operates well by disclosing his requirements in detail. Fourthly, CRM investments pay off handsomely as these clients have long time horizons and high switching costs. Both the client and the consultant invest a lot of resources in the relationship building exercise. The client may find it risky to switch to another consultant and the consultant may find it a major loss to lose the client. Fifthly, it is relatively an easy task for the engineering consultant to implement such a concept as it operates in the B2B market which is more rational than emotional. Sixthly, nothing is more crucial in slow down/recession than to retain clients. Seventhly, ongoing relationships with client is a major business asset as an intimate and delighted client pays less attention to the competitor's offerings. Eighthly, it holds the key to project success because for an engineering consultant, the client is the source of human resource, especially in the execution of mega projects where concurrent design, engineering, plant and equipment supply, civil construction, site erection, testing and commissioning take place. A good/candid and cordial relationship helps the engineering consultant in reading the client's mind well, in understanding project's techno economic requirements, in understanding client's expectations, their constraints, their budget provisions, etc. thereby discharging the responsibilities in a personalized manner. Client's active participation, attitude, disclosure of requirement, furnishing of techno commercial and site information, etc. are critical aspects of quality service and successful project engineering and execution. Without client's involvement, inputs, co-operation and prompt action, the service quality may be far from desirable. Ninthly, the implications of co-existence of service provider (consultant) and service receiver (client) heightens the need further. Tenthly, the firm prefer to buy from firm they like because of their past relationships and experiences. Thus, the saga of CRM in engineering consultancy services emerged.

On-going relationship with client is a business asset. The challenge before engineering consultancy firm is therefore, to build on going relationship with the clients, enjoy their confidence and continued patronage. To measure up, it is imperative for engineering consultants to appoint key account managers (KAM) for the key corporate houses.

12.3 Appointing Key Account Manager (KAM)

Having identified and classified the corporate houses as the key accounts, the next task is to build an ongoing relationship with them. Relationship gets built on the edifice of *trust*. Trust develops out of genuine concern for the other person. In order to create an atmosphere of high concern, the engineering consultancy firm must appoint key account managers (KAM) for the selected corporate houses.

To begin with, the account managers can be appointed for Class- A, Class- B and Class- C client due to initial resource constraints and gradually expanded to encompass other classes and new key clients as identified from time to time. The account manager shall preferably be from middle level grade and selected from various offices of the firm, keeping in view their familiarity with the client, their attitude and aptitude, competence, skills, qualification, etc. The account manager should be technically sound and should have excellent communication skill, negotiating skill, initiative and perseverance. They should have complete understanding of the technology, the service spectrum, the competitor's strengths and weakness, firm's constraints etc. to avoid the pit fall of over commitment. The account manager shall be chosen with a view to serve a client at least for some years. If required, new account manager should be appointed, but well in advance so that there is some time overlap for smooth transition [42].

12.4 Differentiated Offer

By submitting a differentiated offer / proposal, engineering consultant can create an edge for themselves by scoring more technical points. The criteria used to select engineering consultants as mentioned in the terms of reference (TOR) need special attention while preparing the proposals. The offers/ proposal must include cent percent technical compliance with additional features, state of the art technology in design and engineering, memorandum of understanding/ association with technology leaders, proposed work plans, methodology to be adopted, curriculum vitae (CVs) of competent engineers having similar project experience, distinct tasks assigned to various team members, best time schedule, reference list of projects executed in the recent past, experience of working with similar project authorities, commercial offerings, milestone based terms of payment, testimonials like project completion certificate, appreciation letters, awards received, etc. Such testimonials reassure prospective clients about the competence of the engineering consultant and act as a catalyst in differentiating it from its competitors.

12.5 Differentiated Service Delivery

The engineering consultant can differentiate its service delivery in three ways - through people, process and pace. As mentioned earlier, engineers are the face of any engineering consultant and constitute the central force in the marketing mix. An engineering consultant can distinguish itself by deploying competent engineers having strong curriculum vitae (CVs) for the project. Further, it can differentiate itself by adopting a superior service delivery process like designing a superior project execution process, opening well staffed and equipped project offices at sites within the client's premises, carrying out project monitoring activities by using the latest versions of project management soft wares, developing need based project monitoring reports, proper billing schedules, performance guarantee of the system. The services should be delivered as promised, in an environment as expected by the client and its project team, by the engineering consultant who understands the project, who are competent, knowledgeable, reliable, responsive, accessible, courteous, creditable, sensitive and transparent. With regards to pace, an engineering consultant can differentiate itself by adopting shortened project execution than what is prevailing in the industry.

12.6 Solution Selling / System Selling Concept

Marketing engineers of engineering consultancy firms face unique problems while dealing with clients. The reason could be attributed to the **6Os** of the industrial buying process. Industrial buying is unique in itself as it is committee buying based on purchase decision making unit (PDMU) concept, rational buying based on well analyzed facts and figures rather than emotional buying based on impulses, the buyer is not the actual user, every PDMU member considers his requirement as priority, negotiation being an inseparable part etc. The **6 Os** of engineering consultancy services buying process are the **O**rganization making purchase (O_1), the **O**bjective of purchase (O_2), the **O**bject of purchase (O_3), the **O**ccasion of purchase (O_4), the **O**utlet of purchase (O_5) and the **O**peration of purchase (O_6). Buying can be classified according to the client's familiarity with the service it intends to buy. It could be a new task, a straight re-buy or a modified re-buy. If it is a new task, uncertainty may be of the highest degree whereas if it is straight re-buy, uncertainty is almost zero. The industrial customers/clients are technology buyers. They do not buy product/services rather they buy technology or the system. They always look for a distinct solution/ benefit. For example, a power plant never buys a boiler rather it buys steam generation technology using an appropriate firing system to generate a minimum of 'X' tons per hour of steam at 'Y' kg/Sq. cm and T ^{0}C to meet the process and power generation steam demand of the various units of the plant.

To have a distinct edge, engineering consultants should adopt solution selling approach. The marketing team should seek to sell a total solution for problem solving rather than an individual component. System selling begins with a service i.e. advisory or counseling, not a sale. The solution selling/ system selling concept is based on consultative relationships with the customers. It aims at 'do not sell' but 'help the customer/client buy'.

12.7 Industry Analysis

An engineering consultancy firm is the part of an industry where there are many firms engaged in similar businesses, there are many clients seeking similar services but having patronage for someone, there are many firms engaged in substitute products, there are input suppliers, there are firms likely to enter into similar businesses. As such every engineering consultancy firm faces competition either directly or indirectly. Thus, for engineering consultants, the industry and the competitors are vital considerations in making a strategic choice. The industry provides the context in which an engineering consultancy firm operates while competitors vie for the same set of clients by offering more or less identical services. It is quite obvious that any strategic choice cannot be made unless the industry and competition have been analyzed judiciously.

For engineering consultants, the purpose of industry analysis is to determine the industry attractiveness, and to understand the structure and dynamics of the industry with a view to find out the continued relevance to strategic alternatives that are there before them. Porter [43] has made immense contribution to the development of the ideas of industry and competitor analysis and their relevance to the formulation of competitive strategies. He advocated that a structural analysis of industries be made so that a firm is in a better position to identify its strengths and weaknesses for deciding modus operandi. A model has been proposed by him consists of five competitive forces – *rivalry among current competitors, bargaining power of buyers, potential threats from firms which make or offer substitute products or services, bargaining power of suppliers, and potential threat from entry of new firms* – that determine the intensity of industry competition and profitability. **Figure 4** depicts Porter's five forces model of competition.

Using this model of industry-competition, engineering consultants can analyze its critical strengths and weaknesses, its position within the industry, the areas where strategic changes may yield the maximum benefit and the opportunities, minimize the threats and take up the challenges.

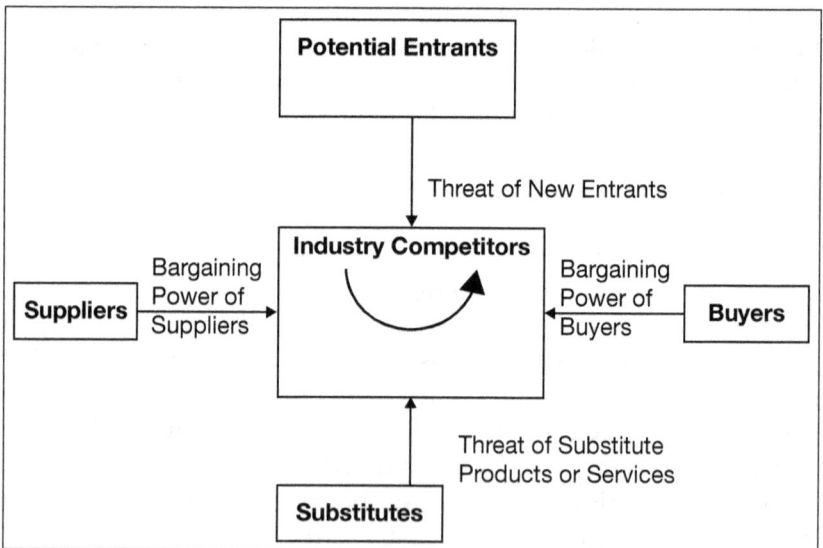

Figure 4: – Porter's Five Forces Model of Competition.

Such an analysis helps an engineering consultant to anticipate changes in the market place, anticipate changes in client buying behavior, anticipate actions of competitors, identify new or potential competitors, learn from the success and failures of others, increase the range and quality of acquisition targets, learn about new technologies, services and processes that affect the businesses, learn about political, legislative or regulatory changes that can influence the business, enter new business, look at own business practices with an open mind, etc.

Porter's five forces model are applicable for analyzing forces of competition in the context of the engineering consultancy industry also. Looking at the global scenario, it can be said that the engineering consultancy industry operates in highly competitive environment. The nature of the competition may undergo significant changes. There are many engineering consultants having core competence in specific sector. Each of them employ their own strategic style to compete in the market. The market is an ephemeral and a constantly changing buyer's market where buyer's have strong bargaining power and hence stiff sales bottlenecks irrespective of their geographical location in the globe.

Some of the clients have set up their own in-house consultancy wings. They import technologies and confine these to their own firms. Substitutes are rare but engineering, procurement, construction and commissioning (EPCC) form of services do offer a threat. Know-how suppliers wield enormous power as sources of technology and basic engineering. Foreign engineering consultants from abroad are potential new entrants for the local consultants. If the investment climate in a country is highly favorable, then foreign engineering consultants will have significant stake in that geographical market. Under the scenario, the engineering consultants can judge their position within the industry and think of the strategic changes that may yield maximum dividend.

12.8 Competitor Analysis and Managing Competition

Competitors are business rivals. Gaining competitive advantage and scoring over competition is the key to business success. The success of any marketing program depends on the strength of competitive analysis on which it is based. Specific competitors must be identified, their strength and weaknesses evaluated and their individual strategies understood. In order to sustain the efficiency in the long run and to differentiate itself from the competitors, the engineering consultant needs to possess a sustainable competitive advantage and competitive orientation.

Poor firms ignore their competitors, average firms copy their competitors and winning firms lead their competitors. Therefore, to lead competitors and have an edge over them, the major ones must be identified, their marketing objectives and strategies, strengths and weaknesses and reaction patterns must be analyzed, competitive intelligence system must be designed and accordingly competitive strategies must be formulated.

According to Porter [44], the purpose of conducting competitor analysis is to determine each competitor's probable reaction to the industry and environmental changes, anticipate the response of each competitor to the likely strategic moves by the other firm and develop a profile of the nature and success of the possible strategic changes each competitor might undertake.

To respond to the competition depends on whether the competitive action has an impact in the market place affecting firm's performance either in the short or the long run. Whether to initiate an action, and how, depends on competitor's ability to respond. The engineering consultants clearly need to assess its competitive advantages and how sustainable they are before deciding how to respond. The next step is an assessment of competitor's strategies and intentions. They must also be cautious of who else might be entering the market, or the likelihood of new competitors. All of this is intended to give a better ability to anticipate the competitor's actions or reactions.

Further, it is not possible for any engineering consultant to tackle all the competitors. They should be selective in tackling their competitors to avoid peril. To tackle competition, they should decide the key competing factors (KCFs) and develop a competitive grid to analyze major competitors i.e. the serious rivals. The criteria used by the clients to evaluate/select/ appoint engineering consultants should be the basic guiding factor for arriving at the KCFs as elaborated above under item **No. 7.0.** The KCFs comparison will give an insight into the competitor's strengths and weaknesses, their strategies, their likely reactions to major business moves, etc. Such an exercise will help in understanding the competitive positions and take necessary actions, preferably offensive.

Porter [44] has opined that most of the firms view competitors as a business threat. While competitors can surely be threats, the right competitors can strengthen rather than weaken a firm's competitive positions. 'Good' competitors can serve a variety of strategic purposes that increase a firm's sustainable competitive advantages and improve the structure of the industry. Accordingly, it is often desirable for a firm to have one or more 'Good' competitors, and even to deliberately forgo market share rather than attempt to capture it.

To tackle competition, engineering consultants should use the key competing factors (KCFs) to go beyond its competitor(s). Quality service, design and engineering credential, past experience, the right set of people, appropriate process, methodology, pace, physical evidences, financial health, fees etc. can be used to preempt competition or create barrier. Otherwise the engineering consultant may join competitors by having a marketing agreement or memorandum of understanding on the division of scope of work to supplement and synergize each other's strengths to reduce competition.

With the above in mind, a competitive grid has been developed for engineering consultancy services considering the relevant KCFs like design and engineering capability, project execution capability, human resources capability, the firm's past experience, financial capability, marketing capability etc. Based on thorough analysis of KCFs, comparative analysis can be carried out for major competitors using the competitive grid. The information obtained from competitive grids are vital and should be used as inputs for the purpose of competitive strategy formulation.

Table 7 presents a competitive grid showing a comparison between an engineering consultant and its prime competitor scored over KCFs on a scale of (1–100). To arrive at the overall competitive position, weighted average method has been used. Depending upon the relative importance, varying weight (illustrative) has been allotted to different KCF. Like, people (engineers) are the most important element in engineering consultancy as proved earlier under item 11.0, maximum weight (say, 50%) can be allotted to this KCFs, followed by design and engineering credential (say, 20%), engineering consultant's past experience (say, 20%) representing the firm's credential and experience, financial capability (say, 5%) and marketing capability (say, 5%). An opinion survey may be conducted among the senior executives of different functional areas of the firm to collect the data on each element of the KCF in a scale of 1-100 to quantify/ arrive at the competitive ranking. Such an analysis makes the situation crystal clear and establishes what image the firm occupies in the market with respect to its range of services vis-à-vis its competitors, i.e. performing better, moderate or lagging behind i.e., whether it is a market leader or market challenger or market follower or market nicher, or a combination of different class of services, and the factors needing immediate attention and augmentation. The competitor's KCFs can be used as guidelines in setting standards or benchmarks and can be used to go beyond. Competitors analysis thus helps in tracking and tackling competition by selecting the right competitors and have a tab on them by formulating the tailor made competitor specific competitive strategy.

Table 7 – Competitive Grid

Key Competing Factors (KCF) (Indicative)	Weight	Scale (1 – 100)				
		Firm	Competitors			
			C_1	C_2		C_N
1. Human Resources Capability (People Factor) a/ Multi disciplinary mix (15%) b/ Qualification and competence (CVs) (35%)	50 %					
Weighted Value (1)						
2. Design and Engineering Capability a/ Work Plan and Methodology (5%) b/ Infra-structure availability (5%) (Technical Information Centre, Archives, Laboratory, CAD Stations, Software etc.) c/ Completion Schedule (5%) d/ Engineering Man Hour Cost (Fees) (5%)	20%					
Weighted Value (2)						
3. Firm's Past Experience a/ Job executed (5%) b/ Value of Single largest job (5%) c/ Typical job executed (5%) d/ Overseas experience (5%)	20%					
Weighted Value (3)						

4. Financial Capability a/ Sales turnover (1%) b/ Solvency (1%) c/ Net Worth (1%) d/ Working Capital (1%) e/ Profit (1%)	05%					
Weighted Value (4)						
5. Marketing Capability a/ Marketing Policy and Plans (1%) b/ Future business procurement strategy (1%) c/ Marketing Tie-ups / MOUs (1%) d/ Net work of offices and strength – Local (1%) e/ Net work of office and strength- Overseas (1%)	05%					
Weighted Value (5)						
Total Score (1+2+3+4+5)	100%					
Competitive Ranking						

Avoiding the Pit falls in Competitor Management

Usually in the engineering consultancy industry, the majority of consultants pay more attention to what competitors have done instead of considering what they might do or will do. There is often an unstated and unrealistic assumption that the competitors will not be aggressive rather than reactive or make their own moves to gain an advantage. Many a times they waste precious resources responding to the wrong rivals. Rather than either lead or follow, it is better to preempt competition, i.e. take action which precludes the competition from taking a similar action.

A point of caution for an engineering consultant is that it they should have a clear competitive strategy. In fact, a firm does not really have a strategy if it performs the same activities as its competitors, only a little better. It is simply operationally more effective. Being operationally more effective is not the same as having the best strategy. Engineering consultants do not always need state-of-the –art tools, huge volume of competitor information, marketing intelligence, armies of expert etc. In B2B marketing, hi-tech is less important than quick actions based on the need of the hour. Engineering consultants must bear in mind that they don't let their rivals gain an obvious lead for long durations. This is because clients make their choices based on what they perceives. Hence, each engineering consultant has something to offer when compared to other available choices. No engineering consultant can compete against all its rivals. The firm should target a few key competitors and ensure success against them. So, the success formula for engineering consultants is to **choose** *competitors, don't let competitors choose them.*

Chapter 13 – Business Procurement Approaches

Scanlon [45] has spelt out the following three approaches which business houses do adopt for order procurements. These are enquiry driven approach, sales driven approach and marketing approach.

13.1 Enquiry Driven Approach

If an engineering consultant waits until the prospective clients identify it as a possible service provider and only responds to the enquiries that result, it has adopted an enquiry driven approach. Here, the engineering consultant waits until the client approaches him indirectly through press notification issuing invitation for bid (IFB) under national / international competitive bidding (NCB/ ICB) or issuing limited tenders and responds to the client's enquiry. The guiding principle here is we are ready when the client requires us. Since such an approach is opportunity based, bid to hit ratio is low as the competitors have also been identified by the client as possible service providers.

Every engineering consultant gets some business in this way but are not sufficient for survival and growth and those who do, are likely to be threatened in the near future. Reliance on this form of business procurement is best described as non marketing. The biggest drawback of this approach is that it reduces client contact to a minima. This develops into a detachment from the market place, which is not compatible with policies to maximize opportunities.

13.2 Sales Driven Approach

When an engineering consultant attempts to render whatever it happens to be capable of rendering, it adopts a sales driven approach. Many engineering consultants see the ultimate ability in marketing as the ability to do just this task. It certainly focuses attention on the immediate situation and will go some way to improve performance. However, its fundamental weakness is that it neglects the question of what clients really want and what they will expect in the near future.

The guiding principle here is that we render what we can offer. It involves offering something assuming that the market / client will accept and then try to defend the service offerings and cajole the customers to acquire it. It is a Win - Lose business approach as it is based on persuasion, compromise or even bargaining. It is battle of service offerings in the market place. Since such an approach protects the interest of the consultant, probability of success is a matter of chance. This approach leads to vulnerability in the long term as fundamental changes in the market requirement take place. This is again a non marketing approach and should be strictly avoided as it is detrimental for long term gains.

13.3 Marketing Approach

When an engineering consultant attempts to identify who are the prospective clients and what they require, approaches them on day one to know their exact requirements, convince prospective clients that the firm is reliable, competent and generally one of the best in its field and is able to tailor services to their individual requirements, negotiates with the client during pre-tendering stage for finalization of the scheme, submits proposals, and then renders services credibly and competitively to their fullest satisfaction, it is said to have adopted a marketing approach. The guiding principle here is that first we identify who are our prospects, what they require (but we never assume them), defend their requirement and then render it using right mixes of the 8Ps. We aim at client delight by selling solutions. It is battle of perception in the mind of the client about the service composition and the consultant offering it. In contrast to the traditional marketing manager's role of finding clients for the firm, the role of the marketing manager of an engineering consultant is to find solutions for the firm's client.

It is a Win - Win business approach as it is based on negotiation and doing justice to the interdependent. Marketing driven approach leads to a direct invitation to submit a proposal, a successful proposal leads to awarding the job to the consultant and a project successfully executed gives more strength for the next round of marketing activity. In a competitive and complex marketing environment, it is the only posture capable of providing long term growth and development. Since such an approach intends to protect the interest of the client, bid to hit ratio is high, return on marketing effort is high, probability of success is almost certain and matter of course. However, simply adopting this approach in principle is not sufficient to ensure future business success. It has to be implemented in the true sense and spirit serving the clients using the 8Ps of marketing mixes and sustaining clients through CRM and KAM.

Chapter 14 – Formulating Marketing Mixes (8Ps) for Engineering Consultancy Services

Once the market has been segmented, target market(s) identified and products positioned, marketing strategy can then address the components of marketing mixes. However, making vainglorious announcement is easy but planning to achieve and really achieving them is difficult. For services industry, it was observed that the traditional marketing mix as coined by Professor Neil Borden (twelve elements) and summed up by McCarthy into the 4Ps namely product, price, place and promotion was inadequate because of three main reasons. The first reason was that the original marketing mix was developed for the manufacturing industries which implied that the services offered by the service companies ought to be changed in a more product like manner so that the existing marketing tools can be applied. This was practically difficult. The second reason was that the marketing practitioners in the service sector found that the marketing mix does not address their needs. They observed that the service has certain basic characteristics which in turn have marketing implications. The third reason was that since services are basically different in comparison to physical products, the marketing concepts have, therefore to be developed in the direction of the service sector. The above three observations suggest that a revised framework for service marketing mix is required and the dimensions of each of the mix elements need to be redefined.

Booms and Bitner [46] have suggested an expanded marketing mix model arising out of the above three observations. Magrath [47] endorsed such an approach and categorically pointed out that when marketing services, the 4Ps are not enough. He said that the other three (3) Ps, strategic elements namely, personnel, physical facilities and process management that capture management's attention, must be included in the marketing mix.

A number of marketing research studies supplement the relevance of each of the 7Ps. Shostack [48] gave due emphasis on physical evidence. Cowell [13] identified the issues in process management. Further, Judd [14] came out with the importance of people. Following the trend, Agrawal [49] expanded the mix of marketing elements to eight variables by including a very important **P**, i.e., **p**ace. Agrawal suggested that the pace of the marketing response is controllable by marketers and therefore is to be included in the marketing response as are the rest of the marketing elements. Also, pace is not only an independent element of the proposed marketing mix, but, is also viewed as the enveloping element for the rest of the seven elements.

Bearing the above in mind, and as mentioned in Chapter 8, engineering consultancy marketing can be defined as the process of identifying clients and their changing techno-economic project investment requirements, converting them into required **p**roducts/ services, making them available at convenient **p**laces, **p**ricing them reasonably (charging fee), communicating (**p**romoting) with them and initiating action (sales), rendering services with **p**ace, execution through proper **p**rocess, **p**eople supported by **p**hysical infrastructure, so that the process of repeat order continues through client delight and competitive differentiation.

In this definition, marketing research assumes the task of identifying clients and their changing techno-economic project requirements, while marketing mixes **(8Ps)** are used to deliver solutions. Thus, engineering consultancy marketing commences much before the services are rendered and continues even after the services are over. To succeed in this marketing warfare, it is essential for engineering consultants to know the markets/ clients, their changing requirements and the competitive forces as well. This calls for market acquaintance, understanding the clients and customization of the services. Thus, engineering consultancy marketing is a three tier system (3S) comprising of searching the market/clients, serving market/clients and sustaining the market/clients through client relationship marketing (CRM) and key account management (KAM).

The market dynamics of engineering consultancy is very peculiar. Growth of engineering consultancy business depends largely on the rate and magnitude of the investment taking place in the various sectors of the industry on account of setting up of new plants/projects, capacity expansion of existing plants/facilities, modernization of existing plants/facilities, augmentation, modification and rehabilitation (AMR) of existing plant units, etc. The industries include mining and mineral beneficiation and transportation through slurry pipeline, manufacturing, ferrous and non-ferrous metals, power generation, transmission and distribution, bulk material handling, hydro engineering projects, oil and natural gas exploration and transportation through cross country pipelines, defense projects, space centres, satellite / rocket launching pads, infrastructures like roads, highways, freeways, and bridges, railways, ports and harbors, airports, the healthcare sector, the hospitality sector, high rise commercial and residential buildings, shopping malls, smart cities, etc. etc. just to name a few. Due to this reason, it is usually said that when large corporate/industrial houses/ project authorities sneeze, engineering consultants catch cold.

The market dynamics/paradigms along with unique features, quality elements, client's concern for quality, client's expectations, benefits sought, positioning elements, people factor, competitive forces, external environmental factors etc. add new dimensions to marketing of engineering consultancy services. Clients want that the marketing personnel should act as a bridge between various departments of both the firms as well as the outside agencies, rather than concentrating on a one to one relationship with a particular user department. For successful marketing of engineering consultancy services, engineering consultants must understand these aspects well and the manner in which they impinge on marketing strategy formulation i.e. **8Ps.** Marketing strategy formulation should address the components of marketing mixes, namely, products/services quality, **p**rice/fees, **p**romotion measures, **p**lacement decisions, **p**eople to be associated in service production and delivery, **p**hysical evidences, **p**rocesses and **p**ace of service production and delivery. When strategies concerning the 8Ps are properly developed, they will produce a synergistic marketing effect and yield the required dividends.

14.1 Products/ Services (P_1)

Products/services are the central force in the marketing mix. At the nucleus of any great brand is a great products/services. To achieve market leadership, firms must offer contemporary products of superior quality that offer unsurpassed consumer value. The ability of the firm to develop a product/service that responds to the techno-economic requirements of the clients is at the heart of Marketing Management. Products/services are developed to feed the requirements of clients and are modified as these requirements change. A firm's identity in the marketplace is established through contemporary products/services. The important aspects that a firm must depend upon are the quality of the product/service, contemporariness, reliability etc.

A service denotes a bundle of features and benefits. The most important issue in service product management is understanding crystal clearly what benefits clients are seeking. The service can be a successful only if there is a synergy between the service product from the client's view point and the consultant's view point. To find this match, marketing manager would have to analyze the service at the following levels: the customer/client benefit concept, the service concept, the service offer, service forms and the service delivery system [50].

The first step in developing service offering is to assess the customer/client benefits i.e. the benefits which the customers/clients are seeking. As Levitt [51] said that people don't buy products, they buy the expectation of benefits. Therefore, it is important to identify what benefits the prospective customers/clients would seek from the service offer. From the view point of engineering consultancy, the benefits which the clients seek from the consultant could be many fold. The consultant provides engineering and technical services but the clients seek an expert opinion; an opinion different from in-house opinion, or vetting of opinion of other consultants, or a report based on which he may take investment decisions, or a report based on which he may approach financial institutions for loans, or a report based on which he may approach other statutory agencies for project clearances, or an augmentation, modification, rehabilitation (AMR) report to overcome typical teething/ operational problems or intend to entrust the entire project execution and commissioning responsibility as the owner's engineers, etc.

Once identified, the customer/client benefit concept is translated into the service concept. Defining the service concept helps in answering the fundamental question: 'What business is the firm in?' It defines the specific benefits which the service offers, say as owner's engineer the consultant renders service on single point responsibility basis from concept to commissioning of the project taking all the responsibilities of project execution and co-ordination between multiple agencies.

Having defined the service concept, the next step is developing the service offer i.e. giving a specific shape and form to the basic service concept. It is concerned with the specific elements (tangibles and intangibles) that will be used to render the consulting service. The service concept has to be translated into core, auxiliary or facilitating and support services with a view to define what the firm would offer to its clients. There is a main or core service and around it are the auxiliary or facilitating services and support services. For example, rendering service on 'single point responsibility' suggests that the engineering consultant offers a bundle of different services. The core service is planning, comprehensive engineering, consultancy and contracting services, supervision of construction and erection etc., whereas the services like, system stabilization, assistance in conducting performance guarantee test for the plant etc. are auxiliary or facilitating services and assistance to the client in obtaining loans from the financial institutions by defending the techno-economic feasibility report (TEFR) or detailed project report (DPR), assistance in obtaining statutory clearances from different government authorities, arranging for customs clearance of the imported equipment, pre dispatch inspection of plant and equipment at manufacturer's works, testing of raw materials etc. are support services. However, in engineering consultancy it may not be possible always to draw a line of distinction between facilitating and support services.

In what form should the services be made available to the clients is another area of concern and decision making. Service forms refer to the various options relating to each service element. The manner in which they are combined gives shape to the service form. Say, should all the activities have a separate contract, or should the engineering activities be charged on lump sum basis and site supervision activities be charged on man month basis or should there be a lump- sum umbrella contract for the total activities.

The service forms are translated into service delivery system which is concerned with the creation and delivery of services using the guidelines built into the service offer. Say, should the basic engineering activities be carried out from the consultant's design office and detailed engineering be done at project site. The main elements in a service delivery system are the people, the process and the physical evidence. The physical evidence components have also been called facilitating goods and support goods. These are the tangible elements of the service and they exert an important influence on the quality of the service as perceived by the clients.

Keeping the above in mind, services shall be planned to fit the needs of the individual client/ project authority and modified as those requirements change. Service planning is concerned with development and renewal of the service spectrum. Decisions with respect to how service features, quality levels, and new service development will be used to satisfy the benefits clients are seeking which must be clearly formulated and integrated with design, engineering, technical and site services.

Since in most of the cases, the product/ service line already exists, hence, service planning exercise gets involved in the task of deciding continuance with the existing services (no change), improving existing services (i.e. service modification), withdrawal of unprofitable services (service elimination), introduction of new services (new service development) and diversification. Service decisions tend to be the riskiest variable in the marketing mix and if ill conceived, may prove to be costly affair. New services introduction are a costly venture, entailing company commitment and risk. Further due to the presence of competent competitors, service acceptance can become difficult and the firm may fail to recover all of its development cost.

14.2 Price/ Fees (P$_2$)

Price (fees) is an element of the marketing mix that earns revenue for the firm, whereas the other elements incur cost. It is not just digits on a tag. Price communicates the firm's intended value positioning of its product/service. A unique products/services can command price premium and reap substantial profit. But the marketing realities may be different and may cause customers/ clients to re-evaluate what they are willing to pay for which compels the firm to carefully review their pricing strategy.

In the engineering consultancy sector, pricing (fees) also plays a vital role. Pricing decisions for engineering consultancy services are particularly important, given the features of engineering consultancy (the elements discussed under Section 4.0) and the critical one being the intangible nature of the service product. If price (fees) is not rational, then irrespective of the lineage of the firm rendering /offering it, it will not click. Like David Ogilvy said, "the customer is not a moron." He knows what is best for his requirement and will settle for nothing less, paying only that much which he deems fit. Proper pricing comes from carefully and consistently managing a myriad of issues. Given the complexities, the key to effective pricing decisions lies in evolving a coherent approach. In engineering consultancy, pricing is largely a situational decision. There is no one best price /fees to charge for a given service.

But every price decision must lie within two limits i.e. lower limit [i.e. cost line, i.e. total cost to be sacrificed to produce/render the service, below which no firm would like to sell] and the upper limit [i.e. market line or value perceived by clients beyond which they are not willing to pay.]

Morris and Calatone [52] have suggested that to introduce creativity, industrial pricing program must encompass the following four key components : objectives, strategy, structure and tactics.

The pricing methods to be used should commence with a review of pricing objectives. The Key Account Manager (KAM) must determine what is he trying to accomplish with this particular price/ fees. The answer/ objective might be sales related, income related, competition related, client related, capacity utilization related, etc. The objectives must be measurable, otherwise, it becomes difficult to determine how well they are being accomplished and whether or not the pricing program is working.

If objectives are the performance levels that the firm wishes to achieve, then strategies represent comprehensive statements regarding how price will be used to accomplish the objectives. A price strategy provides a theme that guides all pricing decisions for a particular service and a particular period. However, as mentioned earlier, every pricing decision should generally fall within the two consideration i.e. consideration of lower price limits (cost to the firm) and the upper price limits (value perceived by the market/clients). Within these two limits/considerations, the price/fees must lie. The actual strategy chosen should be based on a careful evaluation of a number of key factors, both cost based (internal) and market based (external). The underlying philosophy, however, must be that every pricing consideration is examined from the client's perspective. The pricing strategy should be on case to case basis and may fall on any one of the categories depending upon the situation: floor pricing, penetration pricing, parity pricing, premium pricing, price leadership pricing, stay out pricing, bundle pricing, differential pricing, etc.

Once pricing strategy has been decided, its implementation becomes the main concern. Implementation of a strategy requires developing a pricing structure, and then a tactical plan. The pricing structure is concerned about which aspects of each service will be priced, how prices will vary for different customers/clients, what will be the payment terms and conditions etc. A number of managerial questions are to be addressed when establishing the price structure like, should the clients be charged the same per Diem fees/ man hour rate, or should the separate rate be charged, and if yes, then on what basis the difference be established, how often and under what conditions will rate be changed, should clients who value the service more be charged a higher fees than the other clients, what will be the negotiation margin, etc.

Once established, strategies and structures may remain in place but day to day management of price/fees may be required. The day to day management of price/fees focuses alternatively on setting specific price/fees levels and using periodic tactical pricing moves.

Price levels refer to the actual price charged for each service, provision for percentage discounts, going for reimbursement of out of pocket expenses at actual, like tour expenditure (to and fro air fare, boarding; lodging and local transportation charges; etc), incidental charges etc. A point of precaution is that low price benefits clients but normally fails to produce either loyal client or consistent profits. In fact low prices/fees and irrational pricing tactics may antagonize client and contribute to the demise of the firm or force it to go insolvent.

With the above in backdrop, the price/fees to be charged from various clients for different services must be carefully developed. The price levels may require frequent modification in response to change in input cost (man hour cost, activity outsourcing cost, out of pocket expenses, absence fees, etc), competitor tactics, bidding situation, evolving market conditions etc. The ability to manage price levels effectively is heavily dependent on consultant's Key Account Managers (KAM), sense of timing in conjunction with internal factors such as engineering hours cost, actual cash outflows for activities like geo-technical investigation, site survey, testing of raw materials, etc., tours, travels, boarding and lodging, out of pocket expense, margin, duties and tax to be paid to the government regulatory authorities, etc.

The possible fees tactics for design and engineering activities could be lump-sum basis percentage of project cost, direct salary/ pay roll cost plus a percentage (%) mark-up basis and for site supervision services could be per Diem fees/man day or part thereof basis as absence fees, per unit of work basis/item rate basis, etc. and all exclusive of taxes and duties like Goods and Service Tax (GST), which shall be additionally borne and paid by client along with each invoice payment. Further, the lump sum / per Diem fees /man day rates shall remain firm for a period of say, 12 months from the date of issue of the letter of award of contract (LAC) and thereafter subject to escalation @ t% per annum.

While taking tactical decisions, the following additional factors must be kept in mind for quoting the fee : scope of services, technology involved– indigenous or imported, repeat assignment (less engineering to be done) or new assignment (more engineering to be done), expected time schedule, payment terms and conditions, collection mechanism (through Letter of Credit, on-line payment through bank, etc.), similar proposals submitted earlier plus escalation provision, recent orders received for similar jobs, cost of the project, future prospect - likelihood of getting additional jobs, past business relations, client's past payment records, payment risks, client's project seriousness, degree of competition, market trend, say, in some foreign markets quoting less is considered otherwise, speculative and hence, detrimental, market condition- growing/ sluggish, existing or a new area, client wise; service wise; firm's core competence; strengths; weaknesses, engineering and non-engineering facilities provided by the client at the site, method of tender bid evaluation by the client, say CBS (lowest L1 bidder will awarded the job), QCBS (technically qualified but lowest (L1) bidder will awarded the job), QBS (technically highest scored bidder will awarded the job irrespective of the fees), or CQCCBS (combined technical and financial highest scored bidder will be awarded the job) etc., provision to be kept for negotition margin, client's preferences/ penchant client-consultant relationship (on going, intermittent, new), etc.

In determining the exact levels, pricing/fees decision must reflect a variety of pragmatic considerations. Price/fee changes must not come across as arbitrary. The client should sense a degree of consistency in the firm's price/fees levels over time. Consultants must be able to justify, in their own minds, charging prices/fees that are higher or lower than that was previously the case. Otherwise, the firm winds up sending a conflicting signals regarding the value of its services, undermining clients' confidence.

A large number of pricing/fees alternatives are available. The appropriate choice requires considerable creativity and keen insight. Inconsistency between overall marketing strategy and service pricing strategy frequently produces failures in the market place. The engineering consultant that positions itself as a premium quality service provider, but then drops prices/ reduces fees when confronted with competitive pressures, undermines its own market positions, confuses clients, and gives the competitors an opportunity for propaganda.

The above four components of an effective pricing/fees programs are not independent, and should never be approached in an isolated fashion. Rather, they must be closely coordinated, with each element providing direction to the next.

14.3 Promotion (P$_3$)

Contemporary marketing calls for more than developing a quality products/services, pricing them reasonably and making them available at convenient places. No engineering consultant automatically becomes known to the target market merely because it is skilled in its profession. Similarly, a superior service having many "firsts of its kind" or uniqueness does not sell by itself unless its differential merits are communicated and positioned well to the target market. In engineering consultancy, clients are less likely to appoint an engineering consultant without proper information, facts and figures. Hence, there is need for effective communication.

Promotion strategy defines the manner in which a firm communicates with its target market/client. The aim here is to position the differential merits of the product / service and associated offerings in a distinct manner. Engineering consultants must communicate with present and potential stakeholders as clients rely more on subjective impressions.

Promotion Strategy provides the basis for formulating communication mix namely advertising, personal selling, sales promotion, publicity and media selection plans. A good communication mixes should have some combination of personal selling, advertising, sales promotion and publicity. While advertising and leaflet emailing plays vital role in B2C consumer marketing, personal selling, public relations, word of mouth, participation in national and international trade fairs, direct catalogue mailing, etc. play an important role in B2B industrial services marketing. To effectively reach the target market, firms must employ multiple forms of the communication mix. A good promotion mix should have some combinations of the various elements. These elements instead of competing with each other, must complement each other to create a favorable climate for marketing.

In B2B industrial marketing, catalogues are effective and powerful promotion vehicle. Catalogue mailing offers an excellent opportunity to a firm to create marketing leads. Being a natural complement to industrial sales, they are referred to as the firm's silent salesman or salesman in disguise or incognito. As, when the purchasing requirement arises, industrial buyers refer catalogues to gather technical information, clarify doubts, narrow down their technical and functional requirements, screen potential vendors, and invite them for more technical discussions and presentations before taking a final purchase decision. The firm whose catalogues are not readily available with the purchase decision making unit (PDMU) members of the buying firm are usually deprived of the selling/ marketing opportunities.

Because of the catalogue's inherent merits like technical comprehensiveness, ease of reference and its wide acceptance in the industry, catalogues are considered to be a natural complement to the industrial sales calls. They help in creating a favorable climate for personal selling as it lays foundation for marketing prior to sales engineer's call/visit. Despite the dot com boom and social media increasingly becoming a major communication vehicle, catalogues may remain a preferred source of information for the user engineer and purchasing departments. If scientifically designed and properly administered, catalogues readership can be increased, interest in firm's product can be aroused, consultant preference can be created and sales calls can be enhanced. The illustration given below justifies the above hypothesis in the context of B2B industrial marketing.

Illustration

Let us assume that N (say, 1000) catalogues are mailed to the target users in the beginning of the year. Out of that, a constant proportion α (say, 80%) of the catalogues actually survive a given time period t, while the remainder are lost, destroyed, or otherwise rendered unreadable. It is further assumed that each catalogue that survives till the rth time period will reach μ persons (say, 3 readers/ engineers) on the average during that period. Then, around 15,000 PDMU members of the buying firms can be prospected as worked out below.

If we assume that all of the catalogues mailed actually land on the target user's desk, then N catalogues will survive time period t = 0 and will reach $N\mu$ persons. Of the N catalogues surviving time t = 0, only $N\alpha$ will survive period t = 1 and will reach $\mu N\alpha$ persons. Of the $N\alpha$ catalogues surviving time t = 1, only $\alpha(N\alpha) = N\alpha^2$ will survive time t = 2 and will reach $\mu N\alpha^2$ persons. Of the $N\alpha^2$ catalogues surviving time t = 2, only $\alpha(N\alpha^2) = N\alpha^3$ will survive time t = 3 and will reach $\mu N\alpha^3$ persons and so on.

Therefore, the total number of PDMU members C reached by the direct mailing of catalogues would be,

$$C = \mu N + \mu N\alpha + \mu N\alpha^2 + \mu N\alpha^3 + \mu N\alpha^4 + \text{-----------} + \mu N\alpha^{t-1} \qquad (1)$$

Multiplying both side of the equation (1) by α, we get

$$C\alpha = \mu N\alpha + \mu N\alpha^2 + \mu N\alpha^3 + \mu N\alpha^4 + \text{----------} + \mu N\alpha^{t-1} + \mu N\alpha^t \qquad (2)$$

Subtracting equation (2) form equation (1), we get

$$C(1-\alpha) = \mu N - \mu N\alpha^t$$

or,

$$C = \frac{\mu N - \mu N\alpha^t}{(1-\alpha)} \qquad (3a)$$

$$C = \frac{\mu N(1-\alpha^t)}{(1-\alpha)} \qquad (3b)$$

Now, as t goes on increasing, (i.e t tends to infinity), the value of α^t goes on decreasing. So, when t becomes indefinitely large, α^t becomes infinitesimal and can be assumed as zero.

Therefore,

$$C = \frac{\mu N}{(1-\alpha)} \qquad (4)$$

Substituting the value of N = 1000, $\alpha = 0.8$ and $\mu = 3$ in equation (4), we get C = 15,000 i.e. 15,000 PDMU members of the buying firm can be prospected which is 15 times the number of catalogues mailed.

The illustration suggests that in industrial marketing, printed catalogues can act as an effective vehicle to reach a large group of PDMU members that influence buying decisions. Proper management will help in increasing the value of α and μ further, thereby enhancing the reach and the readership rate. Hence, catalogues should be viewed as an effective marketing tool, and not abjured just as an expensive throwaway.

For most firms, the question is not about whether to communicate or not, but knowing about whom to communicate, what to communicate, how to communicate, when to communicate, how often to communicate, etc. Whom to communicate concerns the existing clients and the prospective ones belonging to the upcoming sectors where investment potential is high. The firm should communicate its service capabilities in sectors of its core competence. For such a service and sector which is well known, its aim should be motivating customers to take action. In contrast, for new services and sectors, it should aim at creating awareness about its range of services and past similar credentials.

What to communicate concerns the information about the firm, it's capabilities, engineering achievements, mega project executed, sales performance (sales turnover), financial performance (profit), service extension, technological improvements and changes, diversification, new ventures, memorandum of understanding (MOU) signed with global technology leaders, awards received for engineering excellence, research and development and aspects which affect external stake holder community and the firm's overall social responsibility, the differentiating features i.e., the punch line etc.

How to communicate concerns the communication mix, communication vehicles, selection of the right media vehicle, etc. and when to communicate concerns the right media scheduling. For effective communication, the following promotional measures should be initiated:

- Client visit by Key Account Managers on regular intervals.
- Space advertising in national newspapers and journals, business newspapers and journals, trade journals, industrial directories etc. on regular basis. The firm should advertise in media vehicles like important business newspapers.
- Engaging in social media portals like Facebook, micro-blogging platforms like Twitter, Instagram, LinkedIn, Internet Platforms, Pininterest, You Tube, Snap Chat, Gmail, Yahoo mail, Artificial

Intelligence Marketing (AIM), etc. is the best route to optimize reach as these social media portals are the biggest and unarguably the most powerful social networks and are the perfect online content marketing platforms which are instrumental in developing a strong brand image of the firm and leading towards digital marketing.

- Firms that have a presence on the social media portals are viewed as advanced and technically sophisticated. As their numbers continue to grow, clients perceive the remaining firms that are not on the social media as outdated, anachronistic and unable to keep up with the times.
- Maintaining a Web site on the Internet with all modern features gives increased visibility. The home page is the front door to the World Wide Web house. The latest advances in Web page design, appropriate page title, coining of key words, sufficient pages that may contain relevant information in concise text form, graphics, moving images, animated images with great visual effects, creative uses of colour, artistic composition, sound and sophisticated communication links with interactive capability, reader convenience, banners and buttons, guest page, complaint logging system, online query facility, frequently asked questions (FAQs), etc. with proper Website administration has the potential to make a good impression online. However, it should be kept in mind that Web promotion is an ongoing process, and often does not produce instantaneous results. It may take weeks or even months, before users start flocking into a Website.
- While online promotion is important to draw the online traffic to the social media, mass messaging and Website, offline promotion is equally important to attract those who do not come across or cannot acclimatize the online promotion campaign. Direct mail announcing the new social media portals engagement, spreading around through print and electronic media either dedicated or in conjunction with another ad campaign, embossing social media accounts on the company stationery; business card, company brochures, promotional materials like company gifts, etc. are some of the common ways of offline promotion.
- Direct mail to prospective corporate houses/ clients furnishing corporate brochures, catalogues highlighting range of services, major projects completed in the recent past with its technical details, tie-up arrangements, engineering excellence awards received, letter of appreciation from clients, etc.
- Press releases in national newspapers and journals, business newspapers and journals, trade journals, etc. on regular intervals.
- Organizing press conference: At least two conferences should be organized in a year to give wider publicity to the firm's

achievements and capabilities. The conferences should be addressed by the Chief Executive Officer himself and organized at important cities.

- Chief Executive Officer's interview: The Chief Executive Officer should give interviews to the press at regular intervals.
- Technical presentation to key corporate houses/ clients: To boost marketing, the Key Account Managers (KAM) along with technical personnel should visit the customer's premises and make technical presentation on areas of the clients' interest with solution selling approach.
- Participation in leading national and international trade fairs conducted by trade associations, national and international trade promotion bodies etc.
- Conducting client's conferences in major cities: A series of Compact Disks (CDs) on specific areas may be prepared and displayed at client meets/ seminar.
- Publishing journal specifically for the client.
- Participation in leading national and international seminars and presenting technical papers.
- Sponsoring leading national and international seminars.
- Inviting key clients to visit the firm as a guest to have an on the spot study of the firm's capabilities and facilities. Such invitation letter should be from the Chief Executive Officer.
- Using the existing client as advocate by requesting them to issue 'Project Completion Certificates' appreciating the work done by the engineering consultant and then emailing them to other clients and posting it on social media portals, Website, etc.
- Sales Promotion: In sales promotion, someone must be offered something which is different from the usual terms and conditions surrounding the transaction. In this regard, the engineering consultant may offer its services free of cost for defending the report to financial institutions, agreeing to render services free of cost for minor change in scope or time schedule etc. The objective is to stimulate the recipient's/client's behavior and bring it more into line with the firm's interests.
- Public Relations: Public relation (PR) plays an effective role in motivating others to lend credibility to the firm's proposition. It is a specialized business and has a large role to play that goes beyond maintaining press relations.
- Affiliation with reputed international and national engineering as well as consultancy professional bodies like International Federation of Consulting Engineers (FIDIC); Geneva (Federation Internationale Des Ingenieurs- Conseils), Institute of Civil Engineers (UK), American Society of Mechanical Engineers (ASME), The Minerals

Metals and Materials Society (TMS); USA, ASM International, Consultancy Development Centre, Consulting Engineers Association, International Council of Consultants, Confederation of Industry, Project Management Associates, Association for Iron and Steel Technology (AIST); USA, just to name a few.

Successful business communication needs positive support and lead from the top management. Just to approve or have a policy is not enough. It is important that one of the senior managers and group of managers under him are made responsible for ensuring that the policy is put into practice, the practice are properly sustained, the policy and practice are regularly reviewed, the channels of communications are clearly defined and understood by all involved and concerned.

Engineering consultants may use various media vehicles for promotion, but should rely heavily on personal selling through Key Account Managers. For space advertising and publicity, the important point is the correct selection of media vehicles and media scheduling. The media vehicle chosen must be credible and should enjoy a reputation of being trustworthy. A wrong choice of media vehicle will result in adverse promotion. Media scheduling is important for obtaining maximum impact and reflects the manner in which expenditures are to be distributed.

14.4 Placement (P_4)

Making products/services available at convenient places at the right time is another important key element of marketing success as it avoids stock out or stock piling situation, both being undesirable. The products/services must be delivered to the customer/client when and where they are required i.e., at the right time and the right place. In marketing parlance, it is also called Placement, which deals primarily with channels of distribution.

A channel of distribution is thus a pathway for product to reach the market. A Firm's chosen channel affects all other marketing decisions. The channel of distribution could be a direct one (i.e., from firm to consumer) or an indirect one (i.e. from firm to consumer via intermediaries) or a combination. Selecting appropriate pathway to market can be like crossing a minefield or descending a slippery slope. Going to the market directly and at the same time via intermediaries (marketing representatives wholesalers-dealers-retailers) can prove to be conflict ridden strategy because the firm's intermediaries suddenly see the firm as one of their biggest competitors. A firm who makes an early mess of its distribution channel setup in the chasing process of an emerging market or entering into a new market pays dearly for its mistake as the market matures.

The features of engineering consultancy in general and intangibility and inseparability in particular (as mentioned in Chapter 4.0) make the service delivery channel very short, eliminating use of intermediaries and relying mainly on the direct channel as the only possible way of reaching the clients and serving them effectively. Since direct channel is the only option, therefore location of service centre i.e engineering offices is the first decision variable in planning placement strategy. Since engineering consultancy services are highly interactive based, placement strategy concerns with deployment of engineering manpower at the right place and at the right time to ensure consistency with the client's interaction requirements, project execution methodology and schedule. Place also has importance as the environment in which the service is to be delivered and how it is to be delivered, and is considered to be a part of the perceived value and benefits of the service.

The decision regarding location of service centre, to a large extent depends upon the nature of the engineering consultancy service. In deciding the location of the service centre, the engineering consultants should raise the following questions: what are the trends within the sector, how are competitors reaching the client, what does the client require, what is the degree of interaction required, are accessibility and convenience critical factors in service choice, if service is not provided in a convenient location will the acquisition of the service be postponed, could some competitive advantage be obtained by going against or circumventing the norms, is it technology based or people based, what is the degree of interaction required, etc.

For rendering engineering consultancy services, there are two possibilities: Firstly, during project planning and conceptualization stage, the client is required to visit the consultant's engineering office to discuss various techno financial issues to finalize the project scope. If the interaction required is important due to the nature of the assignment and frequent visits are required by the client, then convenient location of the consultant's engineering office is important. Based on the discussions, the project report is finalized and "Go Ahead" decision is conveyed to the consultant. In such circumstances, the location of the full fledged engineering office in the close proximity of the client is very important and sometimes becomes the reason of patronage. To meet this requirement, consultants must have multi located network of offices at least in the important cities.

Secondly, when the 'Go Ahead' decision is conveyed and front is made available to the consultant, the engineering activities start which is followed by site activities like preparatory work, enabling work, earth work, construction of civil works, supply of plant and equipment, erection of plant units, its testing and commissioning, etc. The consultant is required to undertake project management and supervision of these activities. In such circumstances, the consultant has no discretion as these services are to be provided at the client's premises/ project site. The consultant has to be physically present at the site. To meet this requirement, the consultant has to open a project office at the site and deploy its engineering manpower and resources there for effective co-ordination and supervision.

14.5 People (P$_5$)

The importance of people (engineers) in the marketing of engineering consultancy services marketing has already been established in **item No. 11.2.** In engineering consultancy, people (engineers of various disciplines) constitute the central force in the marketing mix. It is the people (engineer) element, which ultimately makes the difference. The key in quality service lies with the engineers involved in the service delivery system and their passion, verve and zeal for quality. Dedicated and competent engineers of the desired discipline with right traits are crucial. The ability of the firm to deploy the right mix of multi disciplinary engineering manpower that meet customer's project requirement lies at the heart of engineering consultancy marketing. The success of an engineering consultancy firm is tied closely to the management of quality people. An engineering consultancy firm may succeed as a consequence of the effective management of people.

14.6 Physical Evidence (P$_6$)

As a rule, products lack intangibility and services lack tangibility. Services require clues to prove that they exist and in the same form as is being claimed. Physical evidences provide the necessary clues which helps in adding tangibility to the intangibles. For services which are rendered in close association with the client, it is considered to be an integral part of the service offering and the service quality is demonstrated through it. In contrary, an industrial product on the other hand hardly requires evidence, since, it is evident in itself. Further, acquiring of engineering consultancy services entails a high degree of uncertainty and anxiety. In this aspect, physical evidences play a vital role in risk reduction as well in service buying.

Shostack [53] suggested that physical evidence includes two types: essential and peripheral. The essential evidence serves only for the sight of service buyers and cannot be taken away after the use of the service. It represents the service firm's physical infrastructural environment where the services are created, interacted and rendered to the client. The infrastructure available with the engineering consultant like the exterior appearance, aesthetics and interior décor of the engineering office, office layout, extensive in-house computerized data base updated constantly, computing facilities like 1:1 ratio of engineer to CAD machines, latest versions of self developed and procured software, etc., latest communication links like Video Conferencing (VC), E mail, Internet connectivity, phone, electronic gadgets, technical information centre, archives, backup agreement for specialized testing services from agencies like scientific and industrial laboratories reprography facilities, etc. are constituents of the essential physical evidence.

Agrawal [49] has mentioned that the essential evidence is so dominant in its impact on the service purchase and use that it must be considered virtually an element in its own right. These constituents though cannot be possessed by the client but do contribute towards the value enhancement of the service. The common element in these is that they are all controllable and can be used to position the service, build a strong association in the client's mind and also to create service differentiation from the competitors. Every consultants must possess them so that these can be used as extended service proposition (ESP).

Peripheral evidences, unlike essential evidence, can be possessed by the customer and taken away. Agrawal [49] These are those constituents which actually possessed as a part of the purchase of the service. Peripheral evidence plays an emotional role in evaluation of a service by client before, during and after purchase. A nicely structured report or drawing folders having colored graphics, etc. are examples of peripheral evidence. Such evidence has no independent value on its own, but represents a right to experience the service. A report or a drawing folder has no independent value in itself, if it does not contain technical information and various required details with calculations, assumptions, typical sectional views etc. Such evidence 'adds on' to the value enhancement of the output of the essential evidence. Hence these must be designed keeping in mind that such evidences do project the image of the firm.

14.7 Process (P_7)

Process means work activity involved in rendering the services. In a service firm, the system by which services are created, interacted and delivered to the client constitutes the process. Any procedure, task or mechanism which helps to create, interact and deliver service to the client falls under this heading. Process of rendering the services is another major factor within the service marketing mix as the client perceives the delivery system as part of the service itself. For engineering consultancy firm, it is not the marketing, but the service delivery process that also becomes one of the deciding factors for bagging future assignments. The process varies according to the nature of the assignment, scope of the services and the client. The importance of process management is that it assures service availability and consistent quality. As people (engineers) play a critical role in the mix (interactive marketing), hence, they will be severely handicapped if the process is inherently flawed. This means that operations management aspects are of great importance as it would decide how the process of service delivery would function. Without sound process management, quality service is extremely difficult.

Payne [54] suggested that the identification of process management as a separate activity is a prerequisite of service quality improvement. In engineering consultancy, prompt delivery of service involves tasks such as work order acceptance, formation of task force with project leader, arranging kick off meetings, preparation of project management and monitoring schedule/ PERT network, undertaking engineering and site activities as spelt out in the PERT network, deployment of the right mix of manpower to the site, approval of package contractor's billing and shipping schedule, approval of package contractor's drawings, carrying out inspection at contractor's works as well as at contractor's vendor's works, as well as site and issuing inspection certificate (IC) or inspection waiver certificate (IWC), timely certification of progress of contractor's work, issuing material dispatch clearance certificate (MDCC), processing of progress/ milestone invoice of contractors with respect to tenability, submission of payment recommendation to client for release of contractor's invoices of contractor's invoices etc. are outcome of process.

Every engineering department must have clearly laid down procedures for creating and delivering its output. The activities to be carried out for creating and delivering the output must be documented well and shall form a part of the department quality manual (DQM). They must be broken down into logical steps and sequences. Steps which introduce the highest prospect of something going wrong because of personal justification, choice or chance must be identified. Then deviation or tolerance standards for these steps may be set, thereby providing a performance band for functioning. By adopting this approach, errant processes can be made to fulfill their purpose in a more consistent and service enhancing way.

The post contracting phase is the actual phase for keeping the client delighted. After all at this stage, the client physically acquires/ receives the services of the engineering consultant and tries to compare it with their pre-sales expectations as translated/ mentioned in the contract document/ work order/purchase order. The engineering consultant should sincerely try to keep the client sold. Deviations, if any, from presales expectations/ contract document to actual acquisition should be immediately attended to. Post contract activities like opening of full fledged site office; deployment of engineers of required disciplines at site, etc, as per contractual obligation/ commitment, should be seriously carried out in error and omission free manner as this phase is vital for building strong relationships and client patronage.

During the project execution phase, on many occasions, the client's and the engineering consultant's project team have plenty of points to discuss on the variances for smooth implementation of the project so that both the parties can achieve the desired goals. In case of divergence, the engineering consultant may convince the client to refer the work order, kickoff meetings, progress review meetings, and may plead that these services are beyond their scope. However, if the client still insists, the engineering consultant, in the interest of the project may agree to render such services at additional fees or even at no cost, depending upon the nature and magnitude of the requirement on mutually agreed terms. But in no case the engineering consultancy firm should play games with the client and take un-due advantage of the situation. Such a practice should be avoided as it is against the philosophy of CRM and KAM.

Further, engineering consultancy is an intellect and information based service which is concerned with both mental stimulus processing services as well as information processing services. As such, information technology tools should be employed widely in producing and delivering the service. The service output should be created using latest IT hardware and software and they may be transformed into a more enduring and tangible form as reports, drawings, or disks. These outputs can be delivered through e-mails to clients as and when they are required at the site thereby economizing on efforts, time and cost.

14.8 Pace (P_8)

Based on the emerging icon of the new marketing order, time and speed, Agrawal [49] expanded the marketing mix elements to eight variables and proposed a new marketing framework by adding a new dimension 'Pace'. The Net and the Web technology today give very little time to individuals and firms to react and respond. The time has already arrived when every firm has to learn to conduct business operations at the speed of human thought. As such, pace has emerged as a strategic weapon as there is a race against time.

Pace in engineering consultancy means the speed at which the various services are rendered for timely completion and commission the project. Pace has a direct impact on client delight and project economics. Services produced and delivered by an engineering consultant at high pace reduces the gestation period of the project thereby arresting time and cost overruns. The project gets completed before schedule. Completion of facilities before time schedule offers numerous benefits. The important ones being, interest during construction (IDC) on the debt capital getting reduced, expenses during construction (EDC) gets reduced and the plant starts commercial production thereby earning revenue before time, etc. All these lead to higher return on investment (ROI) and improved project economics.

A consultant who is willing to execute the project in time schedule less than their competitors always gets a patronage, even at higher fees, other things being same. Engineering consultants must take this aspect seriously and should voluntarily come out of their own with the best time schedule at the first occurrence to have competitive advantage. To increase pace, the firm should standardize its working, develop detailed project management network, review the project progress, bring the instances of variance, take remedial measures to nullify the causes of variance so that the project gets completed successfully in time. It is thus imperative for engineering consultants to act fast without compromising on quality or cost. In engineering consultancy, delight represents excellence in every respect i.e. all the 8Ps including quickest delivery and prompt response.

14.9 The Need for Integration of the 8Ps

Marketing of engineering consultancy services requires an extended marketing mix comprising the 8Ps. The eight elements of marketing mix should be knit into a distinctive output that delivers service quality. The engineering consultancy firm has to lay great stress on the last four elements of the marketing mix (P_5, P_6, P_7 and P_8) and combine them with the first four (P_1, P_2, P_3 and P_4) to achieve a harmonious blend to fulfill clients' specific expectations. All the 8Ps are important, as changing one element creates an impact on the other. A good fit between the marketing mixes (8Ps) and each client segment and between the marketing mixes (8Ps) and the firm's strategic capabilities must be ensured to reduce the negative influences. It is essential, therefore, that an overall customized marketing mix strategy be developed for different class of clients, Class-A, Class-B, Class-C, others etc., to ensure that all elements are mutually supportive and synergistic. The optimum marketing mix strategy, therefore, involves understanding clients' expectations well, organizing engineering resources, strengths and weaknesses, deciding upon levels of expenditure and being clear about the expected results.

Chapter 15 – Engineering Consultant's Contractual Obligations

The success of a project hinges on the excellence exhibited by the consultant in the project appraisal, conceptualization and planning stage as well as at the execution stage. The consulting engineer should exercise all reasonable skills, care and diligence in rendering the services and should carry out all his responsibilities in accordance with contractual stipulations and recognized standards. Best engineering practices to be followed, accepting full responsibility for work, timely completion of services, meeting contractual commitments and obligations, designer's performance guarantee, operating in a highly ethical manner without getting involved in grey area (confidentiality of drawings and reports, avoiding conflict of interest, avoiding use of clients' resources, respect for the clients' project team, time and money etc.), indemnity against patent / copyright infringement, etc. are important contractual obligations of the engineering consultant. Irrespective of the size of the firm, the consultants should have an ethical and professional standing. It should be independent and should have no corporate, financial or other links with the process or equipment suppliers and the package executing contractors, or other executing agencies. This is essential to ensure impartial technical advice on the project, without pressures or conflict of interests or prejudices.

Chapter 16 – Conclusions

Engineering consultancy is not only about engineering and technological calculations but is also about aspects like formulating marketing mixes, relationship between individuals of the client's project team and the consultant's engineering team, time taken to render the services, the quality of the engineering workforce, CRM, client selectivity, KAM, etc. Marketing of engineering consultancy services requires an extended marketing mix comprising of the 8Ps. Successful consultancy implies rich and diverse engineering, contemporary technology and diligent project experience and sagacious capabilities with reliability, and client confidence. Understanding the client, knowledge about the project, courtesy, responsiveness, credibility and creativity is consultant's core competence, expert man power (engineers) are its pivotal resource and their concern for the client is the major business differentiator. Today, the opportunities lie not only in acquiring a profitable client rather serving, delighting, retaining, sustaining and growing with them. Instead of running after the new ones or everyone, the onus is to retain the profitable ones and run with them. Concern for the client is vital because it is a long duration contact among limited and strategically important industrial corporate houses. Success of the firm is almost single handedly determined by the manner the consultant serves and treats its clients.

The era offers great opportunities but at the same time hurls tremendous challenges. Clearly, the time of making a break with yesterday has come. The engineering consultancy firm must recognize the changing/dynamic paradigms, take up the gauntlet and re-engineer themselves to take on the competition. The new paradigms call for changed organization culture, emphasis on strategic management, multi-disciplinary competence and skill, execution methodology, quality service, client selectivity and ongoing relationship with them, from being essentially concerned with short term transactions to long term relationships, offering services on single point responsibility basis leaving little to be done by the client, meeting stringent techno-commercial stipulations by drawing from the capabilities of the globally best know how suppliers, shortest possible execution time, most competitive fees, etc. In addition to purely technical content, the attention needs to be focused on all these aspects also.

The marketing strategies must be appropriately targeted, and adapted to the particular circumstance and the client. There is a need to transform the commercial image to that of a client caring and quality service provider. The fight is now acquiring the client's mind share rather than mere market share. And, to realize and harness its potential in marketing, engineering consultants must stress on internal marketing. Information technology should be widely used as an enabling tool for carrying out all activities, be it designing the scheme or project management and monitoring or internal communication or external communication. Future of engineering consultancy firms hinges largely upon their ability to reorient themselves to acclimatize to the dynamics of these changes. Receding away from a 'red ocean' and creating a 'blue ocean' experience for the engineers should be the new success formula for them.

References

1. Chatterjee, P. K., and Sharma, A. K., Engineering Consultancy– The New Paradigms, *Technology Trends,* VBS Marketing and Communications Pvt. Limited, Thane, Vol. 2, No. 7, July 2001, pp 57-58.
2. Cowell, Donald W., *The Marketing of Service,* Heinemann, London, 1984, p. 12.
3. Parasuraman, A., Zeithaml, Valarie A and Berry, Leonard L., A Conceptual Model of Service Quality Service and Its Implications for Future Research, *Journal of Marketing,* Vol 49, 1985, pp 41-50.
4. Das, Ranjan, *Strategic Management of Services: Framework and Cases,* Oxford University Press, Delhi, 1997, pp. 5- 6.
5. Carlzon, Jan, *Moment of Truth,* Ballinger Publishing Company, Cambridge, Mass 1987.
6. Gronross, Christian, *Strategic Management and Marketing in the Service Sector,* Marketing Science Institute, Boston, May 1983, Chapter 4, as cited in Leonard L Berry, Valarie A Zeithaml and A. Parasuraman, Quality Counts in Services Tool, in Christopher H. Lovelock, eds, *Managing Services: Marketing, Operations and Human Resources,* Prentice Hall International, London, 1988, p. 217.
7. Zeithaml, Valarie A, Parasuraman, A., and Berry, Leonard L., *Delivering Quality Service: Balancing Customer Perceptions and Expectations,* The Free Press, New York, 1990, pp. 15-34.
8. Parasuraman, A., Zeithaml, Valarie A., and Berry, Leonard L., *SERVQUAL: A Multiple Item Scale for measuring Consumer Perception of Service Quality,* Journal of Retailing 64, 1988, pp. 12-40.
9. Srinivasan, R., *Services Marketing: The Indian Context,* PHI Learning Pvt. Limited, 2012, p 126.
10. Booms, B. H. and Bitner, M. J., Marketing Strategies and Organisation Structure for Service Firms, in J. Donnelly and W. R. George, eds, *Marketing of Services,* American Marketing Association, Chicago,1981.
11. Magrath, A. J., When Marketing Services 4Ps Are Not Enough, *Business Horizons,* Vol. 29, No. 3, May–June 1986, pp. 44-50.

12. Shostack, G. L., Breaking Free From Product Marketing, *Journal of Marketing*, Vol. 41, No. 2, 1977, pp. 73 –80.
13. Cowell, Donald W., *The Marketing of Services*, Heinemann, London, 1984, p. 243.
14. Judd, Vaughan C., Differentiate With the 5th P: People, *Industrial Marketing Management*, Elsevier Science Publishing Co. Inc. New York, Vol. 16, 1987, pp 241-247.
15. Agrawal, M. L., Managing Service Industries In The New Millennium, *Management and Labour Studies*, XLRI, Jamshedpur, India, Vol. 25, No. 2, April 2000, p. 103.
16. Drucker, Peter F., *The Practice of Management*, Harper and Row, Publishers, New York, 1954, p. 37.
17. Levitt, Theodore, Marketing Myopia, *Harvard Business Review*, July–Aug 1960, p. 81.
18. Peters, Thomas J., and Waterman, Robert H., *In Search of Excellence*, Harper and Row Publishers, New York, India Book Distributor, Mumbai, 1982, p. 156.
19. Gummesson, E., Marketing Orientation Revisited - The Crucial Role of the Part Time Marketers", *European Journal of Marketing*, 25 (2), 1991, pp. 60 – 74.
20. Chatterjee, P. K. and Prasad A, Customer Orientation in B2B Marketing - Making it Happen in Engineering Consultancy Sector, *Vikalpa*, Indian Institute of Management, Ahmedabad, Vol. 27, No. 1, 2002, January - March, pp. 35- 43.
21. Sasser, W. E. and Arbeit, S., Selling Jobs In the Service Sector, *Business Horizons*, Vol. 19, No. 3, June 1976, pp. 61-65.
22. Judd, Vaughan C, Differentiate With The 5th P - People, *Industrial Marketing Management*, Elsevier Science Publishing Co. Inc., New York, No. 16, 1987, pp. 241 – 247.
23. Stershic, Sybil F, Internal Marketing- Building Customer Satisfaction from the Inside Out, in Heilbrunn, Jeffrey. eds, *Marketing Encyclopedia : Issues and Trends Shaping the Future*, American Marketing Association, Chicago, NTC Business Books, Illinois, 1995, p. 102.
24. Bartlett, Christopher A, and Ghoshal, Sumantra, Building Competitive Advantage Through People, *MIT Sloan Management Review*, Vol. 43, No. 2, 2002 Winter, pp. 34-41.
25. Parasuraman, A, Services are Performance- Not Manufactured Products, *Strategic Marketing*, January – February, 2004, p 36.
26. Srinivasan R., Services Marketing: The Indian Context, PHI Learning Pvt. Ltd., 2012, p 99.

27. Xavier, M. J., *Strategic Marketing– A Guide for Developing Sustainable Competitive Advantage*, Response Books, New Delhi, 1999, pp. 188-190.
28. Zeithaml, Valarie A. and Bitner Mary J., *Services Marketing-Integrating Customer Focus Across the Firm*, Tata McGraw-Hill, New Delhi, 2001, pp. 15-16, 289.
29. Parasuraman, A., "Services are Performance- Not Manufactured Products", *Strategic Marketing*, January – February, 2004, p 36.
30. Barnes, J. G., The Role of Internal Marketing : If the staff Won't Buy it Why Should the customer ?, *Irish Marketing Review*, Vol 4 (2), 1989, pp. 11-21.
31. Greene, Walter E., Walls, Gart D., Schrest, Larry J., Internal Marketing, the Key to External Marketing Success, *The Journal of Services Marketing*, Vol 8, No. 4, 1984, pp. 5-13.
32. Kimura, Tatsuya, An Empirical Study into the Current State and Structure of Internal Marketing in Japanese Companies, 2011 2nd International Conference on Economics, Business and Management, IPEDR Vol. 22 (2011), IACSIT Press, Singapore, pp.77- 81.
33. Chatterjee, P.K., Strategic Issues in Internal Marketing- A case in Engineering Consultancy, Rai Management Journal, Vol.7, Issue 3, June 2009 p136.
34. Gronross, Christian, A Service Quality Model and its Marketing Implications, *European Journal of Marketing*, Vol. 18, No. 4, 1984, pp. 36-44.
35. *Kotler, Philip and Keller, Kelvin L., Marketing Management (15 e), Pearson Education Limited, Noida, 2017 p.265.*
36. Kotler, Philip, *Kotler on Marketing : How to Create, Win and Dominate Markets*, New York, The Free Press, 1999.
37. Seybold, Patricia, *The Smart Manager, Financial Express*, New Delhi, August 7, 2003.
38. Porter, Michael E. (1980), *Competitive Strategy– Techniques for Analyzing Industries and Competitors*, New York, Free Press.
39. Whitney, J. O. (1996), Strategic Renewal for Business Units, *Harvard Business Review*, Vol. 74, July-August.
40. Chatterjee, P. K. and Prasad, A., Strategic Issues For KAM Implementation in Industrial Marketing, *Vision*, Management Development Institute, Gurgaon, Vol. 7, No. 2, July – December 2003, pp 28-29.
41. Wayland, R. E. & Cole, P. M. (1997), *Customer Connections : New Strategies for Growth*, Boston, Harvard Business School Press.

42. Chatterjee, P. K. and Prasad, A., Strategic Issues For KAM Implementation in Industrial Marketing, *Vision,* Management Development Institute, Gurgaon, Vol. 7, No. 2, July – December 2003, pp 29-30.
43. Porter, Michael E., *Competitive Advantage- Creating and Substituting Superior Performance,* Free Press, New York, 1985, pp. 5, 201.
44. Porter, Michael E., *Competitive Strategy– Techniques for Analysing Industries and Competitors,* Free Press, New York, 1980, p. 47.
45. Scanlon, Brian, *Marketing of Engineering Services,* Thomas Telford, London, 1988, p. 7-9.
46. Booms, B. H., and Bitner, M. J., Marketing Strategies and Organisation Structure for Service Firms, in Donnelly, J. and George, W.R., eds, *Marketing of Services,* American Marketing Association, Chicago,1981.
47. Magrath, A. J., When Marketing Services 4Ps Are Not Enough, Business Horizons, Vol. 29, No. 3, May–June 1986, pp. 44-50.
48. Shostack, G. L., Breaking Free From Product Marketing, *Journal of Marketing,* Vol. 41, No. 2, 1977, pp. 73 –80.
49. Agrawal, M. L., Managing Service Industries In The New Millennium: Evidence is Everything, Management and Labour Studies, XLRI, Jamshedpur, Vol. 25, No. 2, April 2000, pp. 99- 114.
50. Shanker, Ravi, *Services Marketing– The Indian Experience,* South Asia Publications, New Delhi, 1998 pp. 18, 21-22, 32.
51. Levitt, Theodore, *Marketing For Business Growth,* McGraw Hill Inc., New York, 1974, pp. 2-24.
52. Morris, Michael H. and Calantone, Roger J., Four Components of Effective Pricing, *Industrial Marketing Management,* Elsevier Science Publishing Co. Inc., New York, 1990, Vol. 19, pp. 321 –329.
53. Shostack, G. L., Breaking Free From Product Marketing, *Journal of Marketing,* Vol. 41, No. 2, 1977, pp. 73 –80.
54. Payne, Adrian, *The Essence of Services Marketing,* Prentice Hall India Limited, New Delhi, 1995, p. 22, 33, 168.

www.ingramcontent.com/pod-product-compliance
Lightning Source LLC
Chambersburg PA
CBHW070736220326
41598CB00024BA/3438